THE ROOTS OF CHANGE

To my parents, and to Susan, Cameron and Lillian

The Roots of Change

Human behaviour and agricultural evolution in Mali

BRENT M. SIMPSON

INTERMEDIATE TECHNOLOGY PUBLICATIONS 1999

Intermediate Technology Publications Ltd,
103–105 Southampton Row, London, WC1B 4HH, UK

A CIP record of this book is available from the British Library

ISBN 1 85339 468 8

Typeset by Kolam Information Services Pvt. Ltd, Pondicherry, India
Printed in the UK by SRP, Exeter

Contents

List of Figures

List of Tables

Acknowledgements

This book could not have been completed without the assistance of many people and organizations. First and foremost are the hundreds of men and women farmers in the villages of Balanzan, Banamba, Banankoro, Bassian, Dégula, Falan, Fansébougou, Gouani, Kalé, Kanika, Karan, Missira, Samako, Sikoro, Sougoula, Tenemabougou, Tielé, Tinguelé and Tomba, with whom I stayed and carried on long conversations. The contributions made by the many DRSPR/OHV and OHVN field agents with whom I talked deserve special recognition. I am particularly grateful to my driver and friend, Bocar Traoré, for getting us safely through the long hours of road travel and for helping me to see sides of Malian life that I would not have experienced otherwise. A special word of thanks goes to Dennis Bilodeau (USAID/Mali) for his support of this research.

Other individuals who deserve special recognition for their contributions to the field research are: Adama Berthé (AMRAD), my Malian research colleague; Maryjo Arnoldi (Smithsonian Institute) and Juliana Short (Indiana University at Bloomington), who helped me during some particularly sticky periods of data collection; Ulli Helberg (DED), Adrian Gnägi (Suisse Development Co-operation) and Jeff Felton (CLUSA) shared their many insights and experiences with rural development in Mali; William McConnell, William Moseley, and David Midgarden shared freely with me their field observations and research findings; and the Peace Corps Volunteers of Balanzan, Bancoumana, Oueléssébougou and Gouani. A heartfelt thankyou is also due to Birama Diarra (Direction Nationale de la Metéorologie), Makan Fofana (DRSPR/OHV), Stephanie Horowitz (Institute for Development Anthropology), Joe Lauer (Michigan State University Africana Library), Gerry McKiernan (CIKARD), Pierre Rosseau (Auburn University), Catherine Weber (Foster Library, University of Michigan), and Anthony Yeboah (SECID) for providing me with information and helping me to locate the many obscure documents and reports that I requested. I am indebted to Constance McCorkle, Bridget O'Laughlin and Michael Loevinsohn for reviewing portions of this book and making a number of valued suggestions. My thanks also to Karen Shaw (ISS) for helping to produce many of the figures that appear in the text. My sincere apologies to anyone whose name I have skipped over. Although many assisted with this research, the opinions expressed here, and any errors, are mine alone.

Among the many to have contributed to this effort I am particularly indebted to two institutions and three individuals: the Department of Resource Development, Michigan State University, for its support in initiating this research, and the Rural Policy and Project Planning programme, Institute of Social Studies, for providing me with the opportunity to complete the writing of this book; the assistance and encouragement of R. James Bingen (Michigan State University) and David Brokensha in completing this and an earlier draft of this material are deeply appreciated. This book could never have been completed without tremendous sacrifice and effort on the part of my family. Words alone fail to capture the deep sense of gratitude that I feel for my companion and best friend, Susan, for her unwavering support and endless hours put into editing the various drafts of this manuscript.

Finally, it was with immense sorrow that on the day that I turned on my computer to complete the final chapter, I learned of the death of Professor

D.M.Warren. Without his initial encouragement none of what follows would have seen the light of day. His amazing energy and numerous good deeds (mostly untold) will serve as a source of inspiration forever.

List of Abbreviations

AV	Associations Villageoises
AT	animal traction
CDR	complex, diverse and risk-prone
CEC	cation exchange capacity
CFDT	Compagnie Française pour le Développement des Fibres Textiles
CIKARD	Center for Indigenous Knowledge for Agricultural and Rural Development
CLUSA	Cooperative League of the USA
CMDT	Compagnie Malienne pour le Développement des Textiles
CSS	Chef Sous-Secteur
DHV	Development of the Haute Vallée
DRSPR(/OHV)	Département de Recherche sur les Systèmes de Production Rurale/Volet OHV
FF	farmer first
FSR(/E)	farming systems research (/extension)
FAO	Food and Agriculture Organization of the United Nations
GV	Groupements de Vulgarisation
ICRISAT	International Crop Research Institute for the Semi-Arid Tropics
IER	Institut d'Economie Rurale
ISNAR	International Service for National Agricultural Research
ITCZ	intertropical convergence zone
IER	Institut d'Economie Rurale
NARS	national agricultural research system
NGO	non-governmental organization
OACV	Opération Arachide et Cultures Vivrières
ODR	Opération de Développement Rural
OHV	Opération Haute Vallée
OHVN	Office de la Haute Vallée du Niger
PAE	Projet Agro-Ecologie
PNT	Phosphate Naturel de Tilemsi
PRA	participatory rural appraisal
PSPGRN	Programme Système de Production et Gestion des Ressources Naturelles
R&D	research and development
R&E	research and extension
RRA	rapid rural appraisal
SAFGRAD	semi-arid food grain research and development
SMS	subject matter specialists
SPARC	Strengthening Research Planning and Research on Commodities Project
T&V	training and visit
UBT	Unité Bétail Tropical
USAID	United States Agency for International Development

1 Introduction: Towards a better understanding of the local processes of agrarian change

THE INNATE CAPACITY for creative thought and ability to purposefully conceive and undertake desired change are unique features of the human condition. With respect to agriculture, the impact of these intrinsic forces on technical and social evolution is clearly evident in the creation and transformation of specific practices and entire production systems over the course of human history. As central players in this evolutionary drama, individuals and groups of farmers derive much of their motivation to undertake change from sheer necessity, as well as from both the desire to achieve specific gains and a general curiosity about the potential benefits of alternative practices—to which can be added a healthy dose of serendipitous discovery. Viewed over time the state of the art in local agricultural production must be conceived as an evolving set of practices (cf. Niemeijer, 1996), where old techniques are continually being modified, and eventually abandoned, and new ones adopted in order to better fit the shifting set of perceived opportunities and constraints within the producer's various social and physical environments.

While these informal processes of adaptation have been responsible for both the genesis and much of the ensuing development occurring within local agricultural production systems, the twentieth century has witnessed the emerging prominence of a new range of actors—public, private and non-profit—who are explicitly dedicated to influencing the transformation of agricultural practices. Ostensibly, the common goal of these additional actors is to enable and assist farmers in their pursuits. Yet, in practice, each tends to be motivated by different sets of objectives, and approaches problems from radically different perspectives to those of the farmers they are attempting to serve. One result has been that over time, the rate, orientation and scope of change taking place within any location at any point in time has increasingly become a product of both local and external sets of influence.

It is against this backdrop—the fusion of farmers' own innate creativity and the input from various technical and social support services—that this study was undertaken in investigating the processes of local agrarian change among farmers in a large, integrated rural development project, the Office de la Haute Vallée du Niger (OHVN), in south-western Mali (see Figure 1.1). In particular, this study focuses upon improving our understanding of the behaviourial dimensions of farmers' individual and joint actions in the generation, adaptation and spread of new agricultural practices. It is hoped that this improved understanding will contribute to a fundamental reassessment of the importance that these intrinsic human processes have for future agricultural development, as well as the exploration of new areas of complementarity between the various actors involved. The remainder of this chapter introduces the basic themes of what, why and how this study was carried out.

Background to the research

Since the decade of independence in the 1960s, sub-Saharan Africa has increasingly come to be characterized as a region locked in a protracted struggle to meet the basic nutritional and economic needs of its growing population. A review of

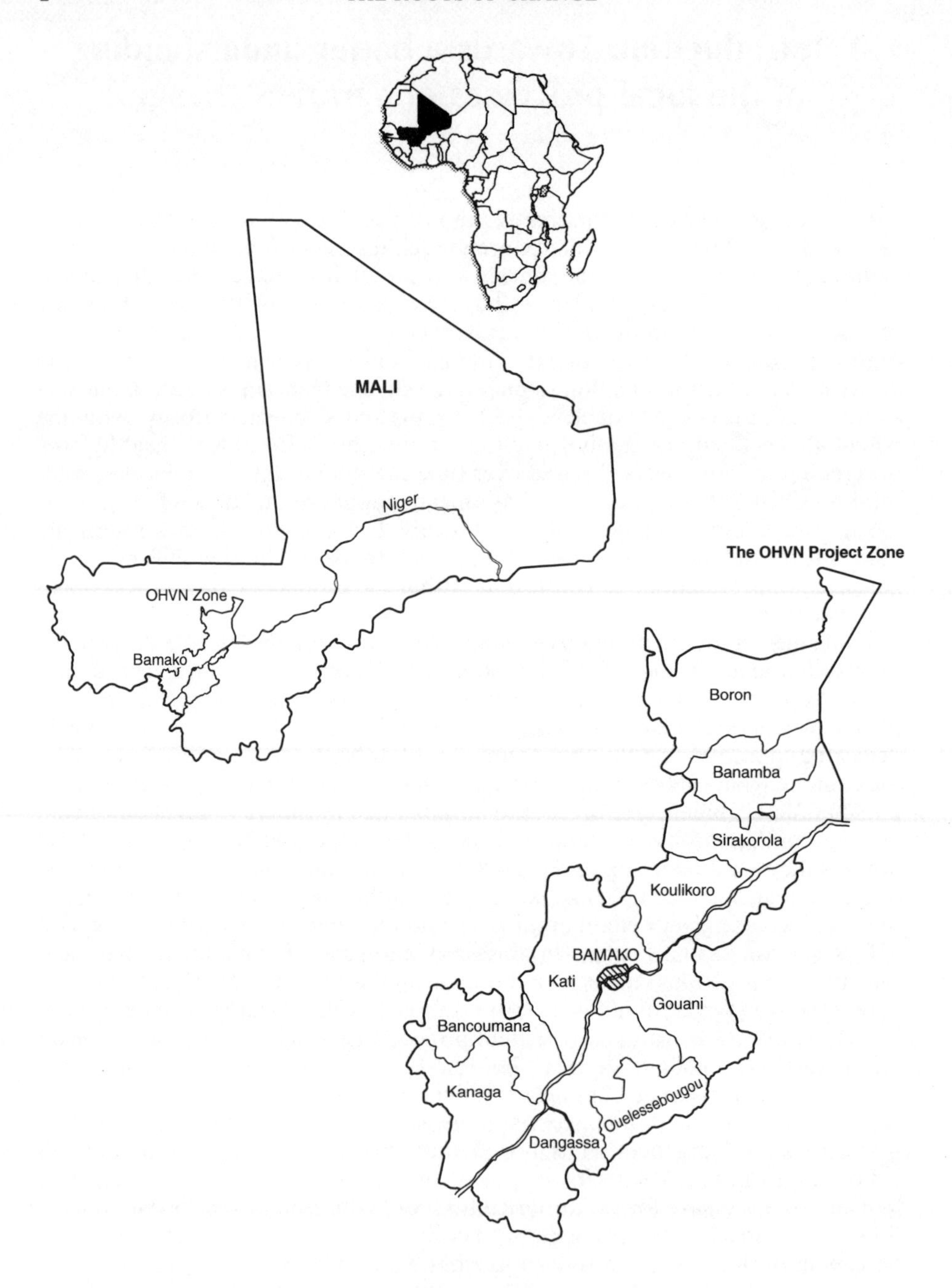

Figure 1.1 *Mali and the location of the Office de la Haute Vallée du Niger*

trends in per capita Gross National Product (GNP) for the region shows that since 1979, few nations have shown positive economic gains, while the vast majority have either just managed to maintain their economic standing, or have declined (World Bank, 1995). The figures on per capita food production

(1979–93) show a particularly worrying trend, with less than a quarter of the countries achieving gains, and over half registering significant declines (World Bank, 1995). Behind these figures, a combination of national policy disincentives, missing or ineffective infrastructure and supportive institutions, lack of trained human capacity, political turmoil, debt burdens, fluctuations in international commodity markets, regional climatic change, inadequate technical packages and approaches, among other factors, have at one time or another been credited with inhibiting a more rapid rate of growth within the agricultural sector (Berry, 1984; Conable, 1991; Diouf, 1990; Eicher, 1989; Wheeler, 1984; World Bank, 1989). In terms of meeting future food security needs, and exploiting the potential that agricultural growth has for contributing to a more general economic transformation, the performances within the agricultural sector over the past several decades have led many to speak of a growing African agricultural crisis (e.g. Glantz, 1987; Platteau, 1990; Timberlake, 1991; World Bank, 1989; among many others; cf. Berry, 1984; and Wiggins, 1995).

In the case of Mali, the pattern of domestic trends that has emerged since Independence in 1960 differs little from that experienced by many other nations during this same period. Two major transitions in government have seen the political-economic leadership of the country shift from a socialist post-independent regime to one based increasingly upon open market principles, struggling to implement macro-economic reforms. As elsewhere in Africa, the past 25 years have also witnessed a steady rise in the annual rate of population growth, which is currently estimated to be between 2.5 and 3.6 per cent (Maiga *et al.*, 1995; Meyer *et al.*, 1993; World Bank, 1995). Although the country as a whole may still lay claim to a land surplus, the open frontier of arable lands is fast closing, and with it the potential of continued expansion-led growth. Already, communities in many parts of the country, especially the more fertile and better watered areas, are beginning to experience shortfalls of suitable land types (e.g. Defoer *et al.*, 1996). The result of increasing land pressure, coupled with major changes in agricultural production practices (from a fallow-based system to continuous cultivation, and a rise in cash-crop cultivation), have led to growing concerns over the level of soil erosion and a long-term depletion of natural soil nutrient stores (e.g. Bishop and Allen, 1989; Day, 1989; van der Pol, 1992). The major (20–30 per cent) reductions in annual rainfall levels, which have affected much of West Africa since the late 1960s (Hulme, 1996), and the cumulative impact of several major droughts occurring during the 1970s and 80s, have served as an additional constraint on farmers' abilities to increase production. Estimates on per capita food production over the fifteen-year period from 1979 to 1993 indicate an overall rate of decline of nearly one per cent per year, while the crash of the world peanut market in the late 1970s, and slump in world cotton prices through the 1980s, have further weakened farmers' abilities to maintain their economic base through agriculture alone.[1] The lack of significant employment opportunities in other sectors of the economy has meant that projected rural poverty levels have continued to rise (IFAD, 1993).[2]

For Mali, as for much of sub-Saharan Africa, the general failure of external investments in rural development to have an appreciable impact on the welfare of the rural poor has been one of the most troubling trends (e.g. CTA, 1996). Particularly disheartening has been the inability of the major investments in the national and international agricultural research systems (NARSs and IARCs) to achieve widespread or sustained successes in alleviating farmers' production constraints and in generating significant new production opportunities outside the high potential areas. Although agricultural technologies represent only one of the ingredients necessary for boosting production levels, the global record of past achievements, current potentials and future needs for technological improvements more than justifies the investment in agricultural research. While collectively African NARSs have had the lowest budgets for agricultural research of any region in the world, the Consultative Group on International Agricultural Research (CGIAR) has invested more in research in Africa than in any other region (CGIAR, 1995; Ravnborg, 1992). In the case of Mali, past national investments and significant donor support have contributed to the development of one of the largest national agricultural research systems in West Africa (ISNAR, 1990).[3] These investments, which have followed a fairly conventional pattern of top-down design and implementation, have achieved only limited successes in stimulating increases in farmers' production levels. Although cotton yields have increased four-fold since Independence and recent gains in irrigated rice production in the Office du Niger have climbed even more dramatically, doubling in the past decade (Maiga *et al.*, 1995), the overall increase in the five-year average cereal yields between 1970–74 and 1990–94 was only 5 per cent, and only one per cent if three-year averages (1970–72 and 1992–94) are examined (World Bank, 1997).[4]

If we shift attention from the national level to that of the OHVN Zone, the conditions encountered are little different. The Zone occupies an area of 31,530km^2 around the capital, Bamako, and informal estimates have placed the rate of annual population growth (reproduction and immigration) within it at above 3 per cent, the highest in the country.[5] To accommodate this growth, the area under crop production has increased nearly 50 per cent over the past decade (Kingsbury *et al.*, 1994). Since 1978, the United States Agency for International Development, one of many donors active in the OHVN, has alone invested nearly $US70 million in an attempt to stimulate agricultural development in the area (USAID, 1978; 1984; 1988; 1993). These investments, which run the gamut of mainstream development approaches—from integrated rural development to farming systems research, and training and visit extension—have provided little opportunity for the involvement of local producers, and at the same time have achieved few significant results in terms of increases in agricultural productivity (Bingen *et al.*, 1994; Kingsbury *et al.*, 1994; Simpson, 1995). The adoption rates of improved varieties and the use of external inputs in food crop production within the OHVN Zone remain low.[6] The increase in production of major cereals has continued to be achieved mainly through the expansion of crop-land under cultivation; however, here too land constraints have become an increasingly common experience for communities in all parts of the Zone. In the early 1990s, millet is the only cereal crop that has shown a significant rise in annual yield levels (averaging 3.5 per cent per annum)(OHVN, 1992a; MDRE, 1997)[7], yet even in this case, there is not a clear link between these increases and the efforts of the formal research and extension programmes, or private sector interests.

Similar patterns of failure in the ability of research investments to generate technological improvements for farmers in marginal environments (as witnessed in the OHVN), have lead development practitioners world-wide to question the efficacy of conventional, 'centralist' models of agricultural development (Biggs, 1989a). In response, a wide range of alternative approaches have emerged since the early 1980s which, for the sake of convenience, are referred to here under the heading of the 'farmer first' (FF) movement (Chambers *et al.*, 1989) to include the many forms of participatory research, farmer-led extension, and a general return to a greater emphasis on local capacity-building. The chronology of experiences and debate that have led up to the current interests in farmers' knowledge, communication and creativity have been well documented (e.g. Bebbington, 1994a; de Boef *et al.*, 1993; Okali *et al.*, 1994a; Sumberg and Okali, 1997; among others), and can generally be regarded as emerging out of the positive experiences involving farmer participation in the on-farm research programmes of early farming systems research initiatives (e.g. Chambers and Ghildyal, 1985; Rhoades and Booth, 1982). Through these early experiences, proponents of the FF movement have come to argue for a number of 'reversals' in the established approaches to agricultural development, in terms of specific practices, professional attitudes and levels of institutionalized support (e.g. Ashby and Sperling, 1995; Pretty and Chambers, 1993; cf. Thompson, 1995). As did some of their predecessors in the colonial era, proponents of the FF movement argue that in order for formal research and extension efforts to effectively address the problems of farmers in 'complex, diverse, risk-prone' (CDR) environments (Chambers *et al.*, 1989), such as the OHVN zone, they must start where farmers are—with their problems, their priorities, their knowledge—and increasingly involve them in the entire development process. The 'complementary methods' advanced by this perspective include making greater use of farmers' knowledge within the formal research programmes, engaging in various forms of farmer–researcher collaboration, as well as promoting the development of farmers' own experimental capacities and information exchange (e.g. Alders *et al.*, 1993; Haverkort *et al.*, 1991; Scarborough, 1996; van Veldhuizen *et al.*, 1997; Warren *et al.*, 1995). This perspective continues to be extended into new directions, and broadened to address new areas of concern (e.g. Scoones and Thompson, 1994; SWI PR/GA, 1997; Sperling and Ashby, forthcoming).

Although the success of alternative approaches associated with the FF movement are predicated on engaging farmers' perspectives, understanding and creative capacities, the internal dynamics driving local agricultural change still remain largely a 'black box'. In recent years, attempts to shed light on these internal processes have led to a flood of papers, books, conferences and workshops about indigenous peoples, their knowledge and the contributions that they have made to local agrarian change.[8] Building upon established anthropological traditions of describing indigenous practices and knowledge within different cultural contexts, researchers have continued to bring to light important elements of local agricultural knowledge (e.g. Bentley, 1992), forms of farmer experimentation (e.g. Potts *et al.*, 1992; Prain *et al.*, forthcoming; Sumberg and Okali, 1997) and the patterns of different types of information exchange (e.g. Mundy and Compton, 1991; Nelson and Hall, 1994). Although these and other studies have advanced considerably our level of understanding regarding specific aspects of farmers' informal Research and Development (R&D) processes, taken by themselves they remain partial and fragmented, separated by differences in socio-cultural contexts, technological themes and the specific feature(s) under investigation (i.e. local knowledge versus experimentation, or information exchange). In order to improve our overall understanding of the behaviourial dynamics of local change (and thereby the ability of new agricultural development approaches to effectively engage farmers, and their skills and knowledge, in development efforts) a more holistic examination will be required of how local knowledge generation and exchange processes function (cf. Okali *et al.*, 1994a). While such a perspective stops far short of suggesting that indigenous capacities will be capable of providing all the answers, it does assert that there is a much greater opportunity (and need) to engage human creativity across a broader spectrum than is currently the case.

The research objectives

In response to the recognized shortcomings of conventional approaches to agricultural development, as well as the fragmented nature of the current understanding and conceptual framework that underlie the alternative FF perspective, this study takes on the challenge of consolidating and expanding our understanding of the internal dynamics of local agrarian change. Using a detailed case history of the evolution of agricultural practices among farmers in south-western Mali, this study examines the processes by which the formal and informal systems of knowledge, communication and innovation have jointly and separately contributed to the recent transformations in local production systems. Based upon field work that was carried out in 1992, as part of a larger study within the OHVN zone,[9] this research began with the basic objective of addressing four general areas of inquiry:

- first, to document the specific changes in agricultural practices that have occurred during the lifetime of the present agricultural managers, and why;
- second, to identify what additional changes these individuals are currently attempting to make, and how;
- third, to identify the major sources, formal or informal, of the information and materials that have contributed to the identified changes, and to explore how these ideas, information and materials have spread between and within communities;
- finally, through an improved understanding of the change process(es), suggest changes in the focus and objectives of formal development activities capable of improving the overall effectiveness of agricultural development efforts undertaken by both farmers and members of the various development-oriented organizations operating within the OHVN.

While this work is explicitly concerned with agricultural change within the context of the OHVN, the lessons drawn from this case-study are clearly applicable to many of Mali's neighbouring countries within the region, and hopefully well beyond. In this regard, the OHVN is particularly well suited as a case-study because of its long history of attempts to stimulate agricultural and rural development through various mainstream development approaches. World-wide, the growing number of local farmer groups that are becoming involved in their own research and information-exchange activities, including those in many non-CDR environments (e.g. The Netherlands, United Kingdom and United States), provides a basis from which generalizations on the patterns and processes of local agrarian change can be drawn at a much broader level.

Theoretical orientation of the study

This research rests upon several theoretical themes and personal perspectives that were important both in framing the original research issues and influencing the orientation taken into the field. The choice of taking a historical perspective in examining local agricultural change was strongly influenced by the compelling accounts written by Knight (1974) and Richards (1985) on the evolution of agricultural systems in East and West Africa, respectively. The remarkable work carried out by de Schlippe (1956) some 40 years ago served to further my interest in using a case-study approach. Combined, these studies, in addition to the findings of a number of field anthropologists and others (e.g. McCorkle *et al.*, 1988; Niamir, 1990; Sharland, 1989), helped to bring to life the notion that local agrarian change constitutes a truly open system involving technological contributions from many sources (i.e. Biggs's (1989a) multiple-source of innovation model), yet one which is ultimately based upon the interpretations, decisions and actions of local producers.

In terms of adaptive processes, the results of a previous study into the social organization of natural-resource management systems in sub-Saharan Africa (Simpson, 1992) led me to view the basic structure and evolutionary nature of change in agricultural practices in terms of the relationships between three overlapping sources of influence: the inter-generational, social reproduction of traditional knowledge; the acquisition of new ideas, information and materials from external actors and agencies (including farmers from other areas, merchants, and contacts with research and extension personnel, and non-governmental representatives, amongst others); and the generation of new knowledge through individual discovery, innovation and adaptive validation. This perspective was further supported by a reflection upon my own experiences in production agriculture, where each of these forces provided a clear source of influence.

Previous field experiences, a general familiarity with the human ecology literature (e.g. Biggs and Clay, 1981; Brokensha and Riley, 1986; de Schlippe, 1956; Fleuret, 1986; Knight, 1974; Western, 1982; cf. Coughenour, 1984; amongst many others) and the opportunity to work on a major study on indigenous African food crops (NRC, 1996; forthcoming) led to considerable attention being directed towards the patterns of diversification in household production activities and the major economic and nutritional contributions of so-called 'minor,' or 'secondary,' crops and products. Tied to this focus was a broader interest in the important relationship between the physical environment and human welfare (e.g. de Schlippe, 1956; Knight, 1980), and in particular, rural people's detailed environmental knowledge and their adaptive capacity to manage an evolving mosaic of productive niches.

Several related issues drawn from the literature and other personal experiences also strongly influenced this study. One in particular, which has increasingly caught the attention of practitioners (e.g. Scoones and Thompson, 1994), concerns the unequal distribution of knowledge, interests and capacities between communities, as well as among individuals within the same community (e.g. Box, 1988; Johnson, 1972; McCorkle, *et al.*, 1988; Swift, 1979). Several studies indicated the existence of local communication channels, or pathways, which influence both the acquisition and local spread of information (e.g. Coughenour and Nazhat, 1985; McCorkle *et al.*, 1988; Nazhat and Coughenour, 1987). Taken together, these observations on the nature of local knowledge systems played an influential role on how field data were to be collected, as well as in the identification of areas which warranted further probing during the actual field work.

Methodological orientation of the study

The overall approach used in the collection and analysis of data presented in this book is based on the seminal work of Glaser and Strauss on the 'discovery' of Grounded Theory. The Grounded Theory approach is an inductive, qualitative research approach developed for the study of complex social phenomena. Overall, the approach is based on the belief that the proper categorization and integration of data will allow the patterns of individual behaviour and social interactions to become visible, and permits the simultaneous proposition and testing of theoretical generalizations (Glaser and Strauss, 1967; Strauss and Corbin, 1990).[10] In operationalizing this belief, the Grounded Theory approach places a strong emphasis on both the use of specific research procedures and the subjective qualities of the individual researcher, who is viewed as the primary research instrument, responsible for carrying out iterative acts of data collection, analysis, categorization, integration and hypothesis-testing. The validity and reliability of conclusions reached under this approach depend not only upon the quality of the data collected, but also on the skill, theoretical sensitivity and creativity of the researcher in critically analysing, organizing and integrating this information into accurate representations of social processes.

Through its explicit focus on the construction of theory, as the 'plausible relationships... among concepts' (Strauss and Corbin, 1994:278), the Grounded Theory approach clearly sets itself apart from other exploratory research approaches. Conceptually, theory developed through the Grounded Theory approach is said to evolve out of 'the continuous interplay between analysis and data collection' (ibid 1994:273), where the emerging theory is shaped by the sensitivity of the researcher, and her or his 'interplay [with] the actors studied', as well as prior knowledge and investigations into the issues under study (ibid 1994:280; Strauss and Corbin, 1990). Pragmatically, theory development is based upon the use of what Glaser and Strauss (1967) term the 'constant comparative method', where new data are analysed and placed into newly created or pre-existing 'conceptual categories' (defined as distinct, self-standing, conceptual abstractions that serve to group and organize the data collected). Through a process termed 'theoretical sampling', categories are developed by using principles of 'minimum' and 'maximum' differences, where in the early stages of research, the researcher seeks data that are fairly homogeneous in nature, or which exhibit 'minimum' amounts of variation. This enables the conceptual properties of the categories, and their theoretical relationships, to emerge more quickly. Once categories are sufficiently developed, the researcher turns to seeking out data for their 'maximum' degree of diversity. The examination and comparison of contrasting or negative cases helps to strengthen and enrich the emerging theoretical connections between the categories. The continuing critical comparison of new data with the existing evidence and categories occurs simultaneously with the generation and testing of hypotheses that explain the relationships between the categories. The use of multiple slices of data, obtained from different sources and through different means, helps to develop the theoretical richness of the categories and their relationships with one another. Glaser and Strauss (1967) suggest that core categories continue to be explored until they become sufficiently 'saturated', or until additional data no longer contribute to their theoretical development; a process, however, that is typically limited by practical constraints of time and access to additional data.

Although proponents of the Grounded Theory approach argue strongly for adherence to the full set of methods, most research situations require adjustments and adaptations in order to fit a broad range of research conditions (Strauss and Corbin, 1990; 1994). In the OHVN case-study, the time limitations under which the study took place, as well as the large size and diverse nature of the study area, required the fusion of Grounded Theory methodologies with various rapid data-collection techniques (see Appendix A). In addition, because logistical limitations precluded multiple visits to each study site, secondary sources of information were used extensively in order to enrich the categories established through the analysis and organization of primary field data. In some instances, conceptual categories developed by other authors in their analysis of empirical evidence were adapted to this study—a practice which, when done with the necessary caution, is consistent with the Grounded Theory approach (Strauss and Corbin, 1990; 1994).

Overall, the Grounded Theory approach was chosen for this study because of its compatibility with the main research questions, data needs and research conditions. In short, it was felt that the approach provided the best fit with the research opportunity, and was capable of producing the type, quantity and quality of information needed for exploring the issues under investigation within the existing time limitations and field conditions. Due to the exploratory nature of the main research issues, the research approach needed to be capable of allowing the lines of inquiry to evolve with the acquisition of new information, rather than generating results that were 'more a product of the methodology', in terms of addressing static questions, 'than of the phenomenon being studied' (Lawler, 1985:3). Second, because local producers themselves hold the bulk of knowledge concerning their practices, the research approach needed to be capable of 'giving voice' to these individuals, allowing them to fully explain their experiences, perceptions and activities, rather than serving as a vehicle for validating the researcher's own hypotheses (Strauss and Corbin, 1990; Marshall and Rossman, 1989). Third, in undertaking this research there was a concern for producing results that could be used as guides for action in making management and policy decisions (Strauss and Corbin, 1994). Although less action-driven than other types of research, proponents of the Grounded Theory approach are deeply concerned with the obligations that researchers have towards society, which carry with them the 'responsibilities to develop or use theory that will have at least some practical applications' (Strauss and Corbin, 1994:281). The ability of Grounded Theory to make such contributions is based on the belief that theory, 'carefully induced from diverse data', will have a high degree of 'fit' regarding its explanatory power over the phenomena which it claims to address (Strauss and Corbin, 1990:23). Finally, the selection of research methods needed to be consistent with the skills and convictions of the researcher who would be conducting the research (Strauss and Corbin, 1990).

Organization of the book

The first half of the book, Chapters 1–3, provides the foundation for the study. Chapter 2 presents a detailed review of the major natural resource systems in the OHVN zone (soils, rainfall, genetic diversity). The physical resources are perceived as not only providing the major parameters within which farmers and external actors must operate, but in a historical sense, as defining the milieu in which local knowledge and farmers' skills have developed, evolved and to which they must continue to respond. In presenting this review, two levels of analysis are pursued. The first concerns the overall system dynamics (i.e. regional climatological system), while the second focuses upon the 'on the ground' realities that these systems generate for local producers. Building upon this description of the physical environment, Chapter 3 describes the highly adaptive nature of farmers' agricultural performances in diverse and risk-prone environments. The chapter then locates the notion of agricultural performances within the larger context of household economic diversification in agricultural, non-agricultural, on- and off-farm activities. The geographic patterns of household diversification are then used to characterize the OHVN zone into different 'household livelihood portfolio' areas, based upon the historical and current influences of physical and institution factors that have been important in shaping the opportunity sets and resulting practices pursued by households in different areas.

The next four chapters in the book (4–7) focus increasingly upon the behaviour of farmers, and other actors, in the maintenance and change of production practices. Chapter 4 explores how farmers' detailed agro-ecological knowledge, informal exchange of information and creative capacities have served as the principal forces in enabling them to keep pace with their rapidly changing physical and social environments. The chapter begins with a description of some important attributes of farmers' environmental and agro-ecological knowledge, followed by an examination of the patterns of local communication and acquisition of materials central to farmer-driven agrarian change. Some basic characteristics, and examples, of farmer experimentation are then discussed. The multiple source of innovation framework is developed in Chapters 5 and 6. Chapter 5 (and the accompanying Appendices B–D) reviews the emergence of and major activities undertaken by the dominant research and extension programmes operating within the OHVN zone. These activities are then analysed in terms of their compatibility and discord with farmers' practices and concerns, and the realities of the physical environment. Chapter 6 presents a summary of farmers' observations on the changes that have occurred in their farming systems since the time of their 'fathers' and 'mothers', involving innovations that have grown out of both formal and informal channels. This review sketches the trends, nature and balance of contributions originating from the principal sources of innovations important to the evolution of household production systems over the past several decades.

The observations on farmers' individual and joint behaviours (contained in Chapters 3, 4 and 6), are reinterpreted in Chapter 7 using a synthesis of postulates and theories drawn from the literature of behaviourial science. Beginning with an overview of the relationships between knowledge, society and individuals, the chapter introduces a basic template for understanding the structure and processes of social differentiation of local knowledge. This overview is followed by an examination of important characteristics of local knowledge, and the different levels, states and forms in which knowledge is found. The chapter concludes by outlining a basic framework for understanding (agri)cultural change. This emphasizes the dynamic interplay between the maintenance of cultural traditions and processes of individual creativity and communication that enable cultures to both sustain themselves and evolve. The final chapter (8) reviews and draws upon lessons learned from the research in making recommendations for strengthening the abilities of formal research and development activities to achieve their desired goals of facilitating agricultural growth within the OHVN zone and similar environments.

2 The physical environment

It is well recognized that farmers throughout the world respond as much to their policy and economic environments as they do to the conditions in their own fields and households when deciding which enterprises to pursue. Yet, once such decisions are made, the ability to produce anything requires a sound knowledge of the natural resource systems that govern an area's productive potential. Because the OHVN zone is not a 'green revolution'-type environment, rich in natural resources, farmers in this and similar areas must be especially adept in using their detailed environmental knowledge in responding to the challenges, manipulating conditions and exploiting whatever opportunities the physical environment offers. For those interested in helping farmers to augment their productivity and well-being in such environments, a basic knowledge of the dynamics of the major natural resource systems that support rural livelihoods is of vital importance. These physical conditions not only constitute an important part of the historical milieu in which farmers' accumulated wisdom and management practices have evolved, but they also define the basic parameters within which the formal research and extension system must operate. In order to conceive and generate technologies and policy interventions relevant to farmers' conditions, formal research and extension efforts must be able to account for the overall diversity and variability of these natural systems in their programming.

The increasingly frequent references made in development literature to the 'diversity and variability' of semi-arid environments, as well as the accompanying 'complexity and riskiness' of the management systems found in these areas, is evidence of a growing recognition of the important differences that these environments possess. Yet, as with so many other development catch-phrases, there is also the danger of these references becoming clichés, serving to desensitize and further distance those not faced with the daily realities of such environments from the harshness and immediacy of the challenges that they impose. One of the aims of this chapter is to prepare the way for (re)introducing a sense of reality into the discussion of farming in semi-arid environments that is to follow in the coming chapters. To do this, two levels of analysis are presented. The first provides a brief introduction to the overall characteristics and dynamics of each resource system (e.g. the climatological system responsible for providing the OHVN and much of West Africa with its rainfall), which is important for understanding their immutable qualities and the inherent limitations they pose for agricultural activities. The second level of analysis involves the translation of the general system qualities into an understanding of the actual 'on the ground' conditions with which farmers in the OHVN, and research and extension programmes, must contend in their management plans and performances, e.g. the diversity of local soils and variability of local rainfall.

Geophysical location

Broadly speaking the natural resource endowments, biological and physical processes and agricultural potential of any region reflects its position on the globe. Occupying an area of 31 530km^2 along both sides of the Niger River in southwestern Mali, the OHVN zone lies near the heart of the West African Sahel. For farmers inhabiting this area, the vagaries of regional rainfall and the general low

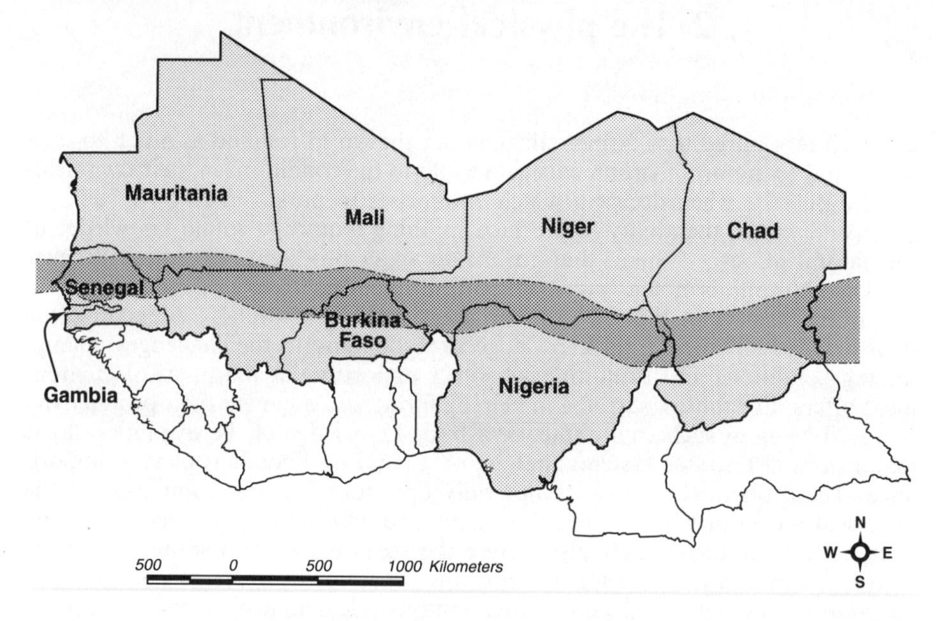

Figure 2.1 *The proximate location of Sahelian West Africa.* This map is based upon FAO data covering the period 1951–1980 with approximate adjustments to reflect the southerly shift of the 200–600mm rainfall zone in the 1980s and 90s. (Source of original map: ICRISAT Sahelian Centre, Niamey, Niger)

fertility of soils present a range and magnitude of challenges that is surpassed by few other regions of the world.

For definitional purposes, West Africa can be thought of as comprising the region roughly bounded by the Atlantic to the south and west, the Sahara to the north, and Lake Chad to the east (Hayward and Oguntoyinbo, 1987). While the eastern border is an arbitrary assignment, that to the north reflects the northernmost advance of the intertropical convergence zone, the region's major climatic force that shapes local agricultural calendars, influences long-term soil formation processes and helps to determine the distribution and composition of vegetative communities. Bisecting this region and spanning the continent from the Atlantic to Indian Oceans, is the Sahelian climatic zone[1] which receives between 250 and 600mm of rainfall annually and is one of four increasingly moist climatological zones that are encountered as one approaches the equator (Saharan, Sahelian, Sudanian and Guinean).[2] Mali contains each of these climatic zones, while the OHVN consists of portions of the southern Sahelian, Sudanian, and northern Guinean zones. Throughout the remainder of this book, the term 'region' will be used in reference to those countries located within West Africa through which the Sahelian zone passes (see Figure 2.1).[3]

Climate

Regional climatic patterns

When describing the climatological conditions of dryland areas to which rural livelihoods and development strategies must adapt, long-term averages are of little value (Nicholson in NRC, 1984b). A few 'wet' years can skew the average well above the mode and medium, and obscure the fact that dry years tend to come in groups (Nicholson, 1982). In addition, such figures fail to reflect the impact that extensive runoff and high potential evapo-transpiration rates, common to such areas, have on crop growth. In reference to the Sahel's climate, variability, punctuated with periodic drought, is a more useful conceptualization of the 'norm' (NRC, 1984b).

The regional weather patterns affecting the OHVN zone are driven by the annual north–south movements of the Intertropical Convergence Zone (ITCZ). Conceptually, the ITCZ constitutes a thermal equator (IUCN, 1989), located between the dry Continental and the moisture-laden Maritime air masses that move north and south in accordance with the earth's position relative to the sun. This north–south movement in the ITCZ drives the region's climate and is responsible for its rainfall. During the year, mean temperatures increase dramatically prior to the onset of the rainy season, with a lesser peak after the rains, followed by a cool-dry winter. During the dry period, the dust-laden *harmattan* winds blowing out of the north-east dominate the landscape, both figuratively and often literally.

The region's single rainy season, which occurs in the OHVN zone roughly between June and September, results from convective activity initiated by moist air being drawn northward off the Gulf of Guinea with the advance of the ITCZ.[4] As shown in Figure 2.2, the northward advance of the ITCZ is characterized by major daily advances and retreats in its position (involving shifts of as much as 12 degrees of latitude in a single day) (Hayward and Oguntoyinbo, 1987). This erratic movement of the ITCZ is responsible for the uncertainty in the arrival

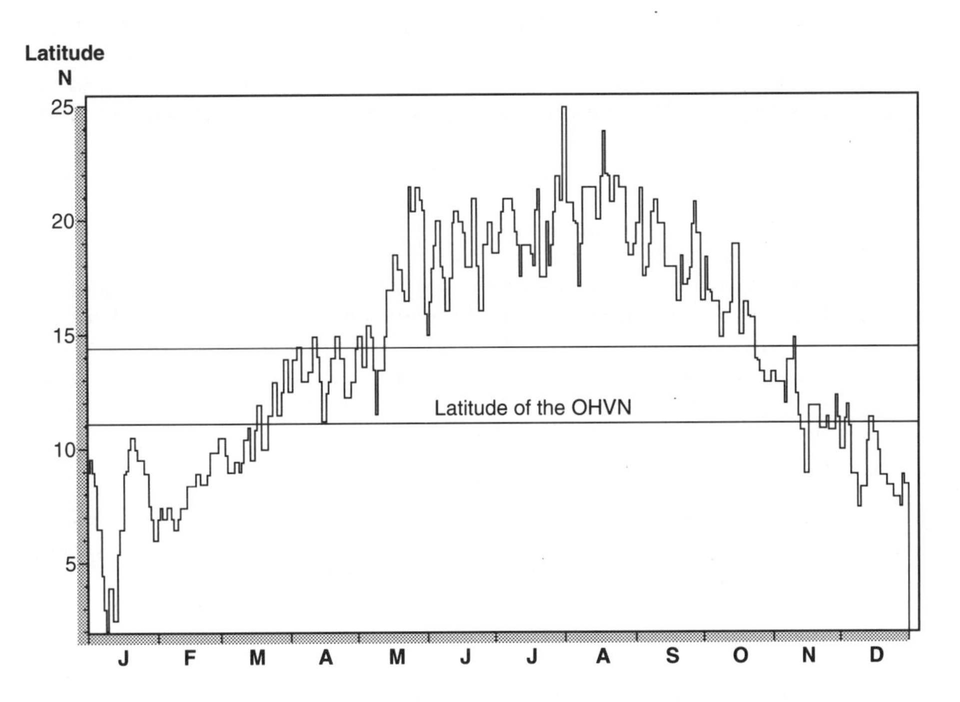

Figure 2.2 *Daily movements in the ITCZ.* (*Source: Hayward and Oguntoyinbo, 1987*)

and length of the rainy season for any one location.[5] As one moves northward, the irregularity in the onset and duration of the rains, as well as the overall quantity of the resulting precipitation generated by this weather system, increases steadily. The coefficient of variability for annual rainfall[6] along this south–north gradient is approximately 15–20 per cent for the northern Guinean, 20–30 per cent for the Sudanian and 30–50 per cent for the Sahelian zone (Nicholson, 1982).

Impact of local climatic patterns

Combined with the temporal uncertainty in the arrival and duration of the rainy season, the nature of individual storm-event formation, discharge and movement accounts for the pattern of high spatial variability in the rainfall experienced on the ground. By for the greater part of the precipitation falling in the Sahel is generated by narrow bands of rainfall, line squalls, embedded in larger cloud clusters that form along the ITCZ (Nicholson, 1982). On the whole, convective activity is generally limited to 1–10 per cent of the ITCZ at any one time, leaving large gaps between cloud clusters, and even larger gaps between the rain-producing line squalls (Farmer and Wigley, 1985).[7] Once formed, line squalls show a tremendous variation in their shape and movements.[8] The leading edge of these storm events typically exhibit intense rates of discharge (see Figure 2.3), with brief periods of rainfall routinely exceeding 100mm/hr (Jones and Wild, 1975; Sivakumar, 1989). This intense rate of rainfall causes line squalls to continually dissipate and expand, with the loss and addition of new moisture in their passage over the landscape. The net result of such widely spaced, constantly evolving storm events is that contiguous land areas may receive widely disparate levels of precipitation.[9] Village *animateurs*, responsible for monitoring rain gauges installed by different development programmes within the OHVN zone, report variations in rainfall of as much as 10mm for single storm events between different gauges located within the same village. Year to year differences in the temporal and spatial distribution of rainfall are even more dramatic (see ORSTOM, 1974).

From the perspective of household decision-makers, whose different fields may be separated by several kilometres, the high degree of variance in rainfall alone requires that each field be managed on an individual basis. This variance is especially prevalent at the onset of the rainy season, when the irregular rainfall is aggravated by the final ravages of the *harmattan* winds, which can raise the already high evapo-transpiration rates by eight mm/day (Kowal and Kassam, 1978). This situation can be further exacerbated by the tendency of some soil types found within the OHVN zone to form surface crusts after the initial rains, thus greatly reducing the infiltration of subsequent rainfall. Depending upon the topography and soil characteristics, the intensity of rainfall generated by line squalls can produce rates of surface runoff that exceed 60 per cent (Rodier, 1982), leading not only to poor infiltration, but also to serious incidence of soil erosion (e.g. Roose, 1977).[10] In summarizing these conditions, Charreau observes that 'dry tropical climates have 6–10 times more erosive power than temperate climates' (1974: 74; see also Kowal and Kassam, 1978).

Recent trends

Since the late 1960s, regional trends in annual precipitation levels have shown a significant decline. Following the relatively 'wet' decade of the 1950s, with many

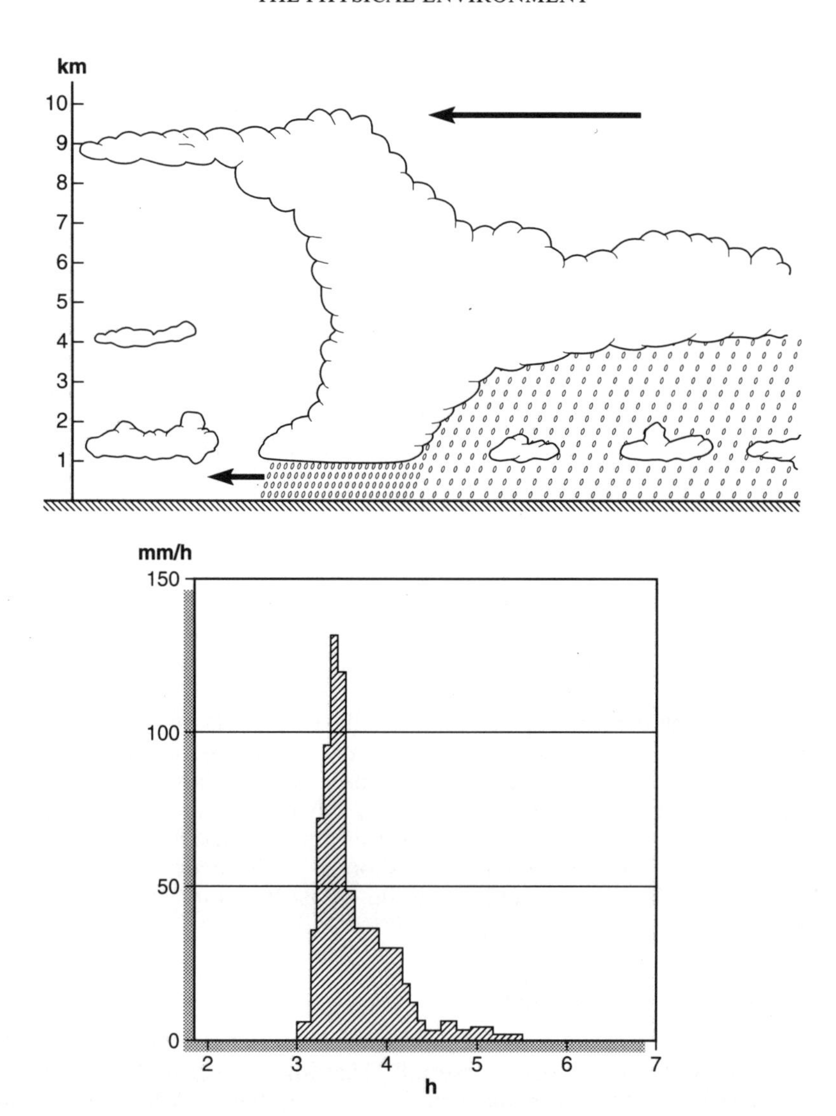

Figure 2.3 *Anatomy of a line squall and discharge rates. (Source: Cochemé and Franquin, 1967)*

stations recording rainfall 60 per cent above their previous average (Nicholson, 1982), the region experienced years of declining rainfall, culminating in the disastrous 1968–73 drought, when much of the Sahelian and Sudanian zones received only half of the rainfall of the 1950s (Nicholson, 1982).[11] Since that time, annual rainfall has yet to return to its pre-drought levels (see Figure 2.4) (Hulme, 1996).

Regionally, the approximate 200–300mm per year decline in rainfall represents a 20–30 per cent departure from pre-drought levels, and corresponds roughly to a 150–250km shift southward in the rainfall isohyets (Hulme, 1996; NOAA, 1979).[12] The reduced rainfall not only limits farmers' productive opportunities,

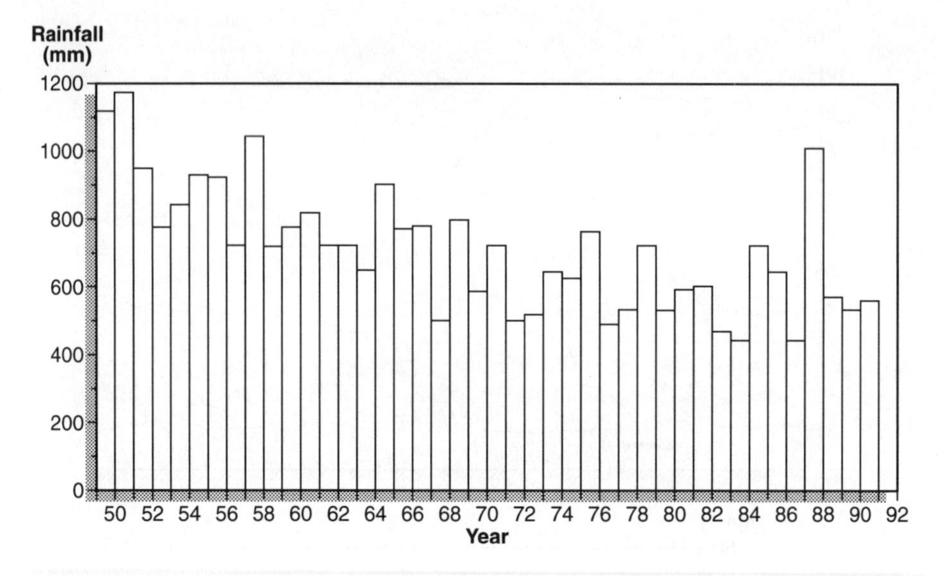

Figure 2.4 *Annual rainfall levels, Banamba Station, OHVN, Mali (1950–1991).* (*Source*: *OHVN, 1992a; SRCVO, 1989).*

effectively reducing the growing season by 25 days, or by as much as 20 per cent (IUCN, 1989), but as noted, lower rainfall levels are also associated with higher levels of annual variability, especially in the onset and length of the rainy season. This increased variability is most profoundly felt during the critical period of planting, where the flush of nutrients released by the first rains makes timely planting essential.[13] Yet, even relatively short periods of drought (which are common during these early rains) can be disastrous for vulnerable, newly germinated crops.

Ground and surface water resources

The region's limited surface and ground water supplies have also been significantly affected by the downturn in rainfall. The Niger River and its tributaries constitute nearly all of the permanent surface water in the OHVN zone. Seasonal stream flows, the extent of river flood levels and inundated areas (*bas fonds*), and the recharge of underground aquifers, are all dependent upon seasonal rainfall. With the decrease in rainfall, farmers in many areas have had to adopt new production strategies, especially those in the southern *secteurs* producing rice through the use of semi-controlled (e.g. McIntire, 1981), natural flood (*crue*) or recessional (*décrue*) agriculture (Harlan and Pasquereau, 1969). For much of the zone, ground water levels have also been negatively affected by the decrease in precipitation. Even during periods of abundant rainfall, neither the extensive aquifer found in most locations at between two and 35m depth, nor the deeper aquifers located in faults in the bedrock of the southern parts of the zone, have enough reserve to support widespread irrigation (PIRT, 1989b; Shaikh *et al.*, 1988b). Villagers note that during years of poor rainfall, certain wells in most villages run dry, and in the northern areas many of the shallow wells constructed to support irrigation of dry-season gardens and the watering of livestock have been abandoned altogether. Widespread, drought-induced mortalities are visible in the tree populations of the northern-most *secteur*, in part because ground water levels have fallen as much as 75 feet in some areas (Shaikh *et al.*, 1988a).[14]

Soil resources

Macro-variations in soils

Next to rainfall, soil fertility is the most important factor limiting crop production in much of the region. When considering the OHVN zone, it is immediately evident that the main agricultural area straddles a distinct north–south division in major soil groups. This division is the product of two major geologic formations (Ahn, 1969) which, through the long-term effects of secondary weathering, account for much of the present-day topography and macro-variations in soil types found within the OHVN. Soils in the southern *secteurs* of the zone[15] are the product of a major up-folding of an extremely old, and generally infertile, bedrock formation referred to by geologists as 'basement complex' material (Ahn, 1969; Charreau, 1974; PIRT, 1989b). Intensive weathering of this pre-Cambrian formation led to the formation of a massive laterite sheet covering much of the zone, which has long since degraded into the broad, crumbling, ironstone-capped plateaus and undulating topography which characterize the area. Scattered inselberg formations,[16] and the prominent escarpment and mountainous area of the Mandingue Plateau along the western bank of the Niger River, provide additional relief. The moderate moisture regime of the area supports abundant plant growth, important in the development of soil organic matter, yet is also responsible for leaching most of the soluble nutrients and clays from the soil profile. As a result, the majority of nutrients available for crop growth are contained in the soil organic matter, which rapidly declines through erosion losses and increased oxidation once the protective vegetative cover is removed and the unprotected soils are subjected to annual cultivation (Jones and Wild, 1975).

In the northern area of the OHVN zone,[17] the predominantly sandy soils have developed from extensive sedimentary deposits heavily weathered during the Quaternary period (Ahn, 1969; Charreau, 1974). As in the South, widespread hardpan formations have degenerated into crumbling plateaus, blocks and beds of plinthite pebbles (Ahn, 1969; Charreau, 1974) which, with the eroded hillocks and broad sand plains cross-cut by dry-wash stream beds, are characteristic features of the landscape. In the far northern reaches of the zone, ancient, fossilized sand dunes are visible. This series of NE–SW facing dunes have long since been stabilized by vegetation (Charreau, 1974; PIRT, 1989b) and is characterized by its own unique hydrological and pedological processes. The lower levels of rainfall and moisture-holding capacity of the soils in the northern area limits production of above-ground biomass, and consequently affects topsoil development. Yet the limited rainfall has also allowed these soils to retain a larger share of their initial nutrient stores in the upper soil horizons through lower rates of leaching (Jones and Wild, 1975).[18]

Micro-variations in soils

Regardless of their origin and composition, the soils found locally in both the southern and northern areas of the OHVN zone exhibit a number of important forms of variations to which farmers respond through their management practices. Such variations are both systematic and random in their nature. The concept of *catena* (the Latin word for chain) (Milne, 1935) is commonly used to explain the systematic transition in soil types 'encountered as one goes from hill top to valley bottom' (Ahn, 1969: 60), due to influences of topography, hydrology, changes in parent material and vegetative communities.

In contrast, differential rates of wind and water erosion across seemingly uniform surfaces, the impact of specific biological features, irregularities in the surface topography and underlying hydrology provide examples of some of the different sources of random variance that influence the productive potential of specific sites, yet which cannot be generalized or accounted for in any systematic fashion (Wilding and Drees, 1978). Many of the sources of this type of random variation are linked to biogenetic forces (Brouwer *et al.*, 1993; Moormann and Kang, 1978;) e.g. differences in vegetative cover; the effects of individual trees on soil organic-matter content and micro-climatic influences (Charreau and Vidal, 1965); the activities of termite communities, which increase drainage and bring to the surface fine, often nutrient-rich materials, including clays (Warner, 1991);[19] and human disturbances in the form of mixing soil layers, disruption of structural integrity, changes in vegetative cover, accelerated (and abated) erosion levels and the removal and addition of organic matter—all of which play a significant role in the formation and evolution of micro-variations in soil structure and fertility. Both the systematic and more random features of soil micro-variation challenge farmers in their ability to identify and respond to the difficulties and opportunities presented by the different production environments (e.g., Brouwer *et al.*, 1993; Carter and Murwira, 1995; Lamers *et al.*, 1995; Scott-Wendt *et al.*, 1988; cf. Mulla, 1989).

Vegetative cover

Large-scale categorizations

In addition to rainfall and soils, vegetative communities comprise a third major resource stream upon which farmers' livelihood strategies depend. Since the initial classification of West African vegetative zones (Guinean, Sudanian and Sahelian zones) proposed by Chevalier (1900), a number of schema and descriptors have been used to differentiate the relatively seamless transition from open desert to closed canopy moist forest (see Lawson, 1986). Overall, however, the composition and location of plant communities within the OHVN zone can be thought of as resulting from the combined influence of two main forces: those associated with the 'natural site', or basic growing requirements of each species, and those of the 'socio-economic site' which, through various human activities, delimit the range where species are allowed to grow (von Maydell, 1990). In semi-arid environments, the natural site of different species is principally governed by water availability—the quantity and distribution of rainfall, topographical concentration of seasonal flows, proximity to ground water, the water-holding capacity of different soils and the ability of species to endure annual, as well as longer-term periods of drought (Bille, 1977; Hills and Randall, 1968; Lawson, 1986). While some species exhibit a great deal of elasticity in their overall requirements, others are much less tolerant (von Maydell, 1990). Variations in soil properties (nutrient stores, pH) (Hills and Randall, 1968), sensitivity to varying levels of sunlight, shade, interspecies allelopathic reactions, grazing pressure, fire and wind are additional parameters that shape the positioning and growth of individuals and entire communities within different sites across the landscape.

A common characteristic of species inhabiting semi-arid environments is their extensive underground nutrient stores that allow them to endure significant and lasting shortages of rainfall.[20] In general, the vegetative communities of the Sahel (both living and represented in the soil seed bank) are very resilient. Observations in the OHVN and several studies show that, even on the most degenerated sites, a significant 'rebirth' of vegetative communities is possible if they are left undisturbed (e.g. Hopkins, 1965; Bradley, 1977).

Because of long-term human influences within the Sudanian and Sahelian zones of West Africa, few undisturbed vegetative communities remain. Historical evidence indicates that fire has been a major factor shaping the Sudanian and Guinean zones for possibly the past 2500 years (Müller, 1855). In some areas the prolonged pressure exerted by grazing livestock has meant a reduction in fragile and more desirable species, causing a shift towards plant communities of hardier and less palatable species. In addition to brush fires and grazing, the cutting of trees for timber, fuelwood and charcoal manufacture, removal of tree stumps to facilitate ploughing, and over-harvesting of secondary products are among the many influences identified as contributing to the general decrease in species diversity and reduction in the overall density of above-ground vegetative cover in the region (e.g. Gritzner, 1988; NRC, 1984b). While recent research in the more humid regions of the South argues that, contrary to popular belief, the forest cover in this area has actually increased over the past century (Fairhead and Leach, 1996), the extent to which this may be true for the more arid environments of the OHVN is probably limited. Nevertheless, there can be no doubt of the importance of vegetative communities to rural livelihoods in the OHVN, ranging from the provision of fuelwood and construction material, to foodstuff, fodder, medicines and numerous other 'secondary' products.

The distribution of vegetative communities within the OHVN

At a minimum, perennial species inhabiting the northern Guinean zone must be able to endure the eight-month dry season between rains, which bring 900–1250mm of precipitation each year. Expanses of open canopy dry forest are common, with patches of closed canopy forest, especially along riverine and undisturbed low-lying areas. The primarily deciduous tree species allow a grass understorey to exist in most areas. This grass cover varies from scattered tussocks in more heavily forested areas, to tall grass savannah species (> 80cm) in the open areas of grassland. Plant communities in this zone are thought to be fire climax, and are the basis for the common reference to this area as 'derived savannah', evident in the abrupt transition from savannah to closed-canopy moist forest (occurring less than 150 metres further to the south) (Hopkins, 1965). In the absence of fire and other forms of human disturbance, it is thought that large areas of this zone, as well as much of the Sudanian zone, would revert to closed-canopy dry forest (Lawson, 1986; cf. Fairhead and Leach, 1996). In agricultural areas, trees with economic importance, dominated by *karité* (*Vitellaria paradoxa*), form the extensive 'agroforestry parklands' for which much of West Africa is noted.

In the somewhat drier Sudanian zone, averaging 600–900mm of rainfall per year, the impact of a historically higher population density,[21] with its associated agriculture, grazing and use of fire, is much more apparent in the structure of woody perennial populations. Save for riverine and protected areas, government reserves and sacred groves, mature closed-canopy forest is less abundant, although dense tree stands of younger trees and thickets are common. The wooded savannah and open glades are populated with both shorter-stemmed grasses and broad and fine leaf deciduous species, characterized by their thick, fire-resistant bark. As with the Northern Guinean zone, the dominant woody perennial in highly populated areas are those with economic value (such as *karité*, tamarind (*Tamarindus indica*) and *néré* (*Parkia biglobosa*), although their densities decline noticeably as one moves northwards.

Averaging between 250 and 600mm of rainfall per year, the Sahelian zone supports both lower human population levels and plant densities. Just below the northern limits of this zone, woody perennials are able to obtain sufficient moisture to spread out beyond the seasonal water courses, reaching densities of up to 40 per cent coverage in some of the better sites of the southern Sahel (Steentoft, 1988). Following a north–south transect, species diversity increases sharply, from just over 30 species in the extreme north, to over 120 species in the South (von Maydell, 1990). This area is dominated by a variety of thorn-bearing acacia and other species that lend themselves to the description of 'thorn-scrub.' Due to the lack of sufficient fuel load to support brush fires (Lawson, 1986), fire is less of a factor in shaping the structure of plant communities in this area. However, intensive grazing and the lopping of branches by herders, especially along transit routes, has resulted in wide expanses of exposed soils and the characteristically architectured shrubs and trees.

Summary

The evolving and continuously interacting natural resource systems described in this chapter present farmers with a dynamic mosaic of resources with which they interact and attempt to manage. Certain elements of these systems, such as soil nutrient levels, are relatively stable, showing moderate, discernable year-to-year changes that are open to management interventions. Other variables, such as yearly rainfall (its onset, duration, and spatial and temporal distribution) and the associated flood levels, offer less scope for interventions and must largely be anticipated and reacted to within the context of a particular season. The composition of the different vegetative communities, and changes in weed populations, provide yet another source of variation that reflects elements of both long- and short-term environmental change, as well as the influence of human activities.

In sum, farmers living in areas at the margin of secure agriculture face greater extremes in environmental variability, and must contend with an overall larger range of contingencies, (and do so with fewer resources), than is the case for their counterparts living in more stable and well-endowed regions. As the following chapters will illustrate, the lack of abundant and stable natural resource systems requires farmers to rely upon their detailed knowledge of local environmental variations and adaptive creativity to secure their livelihoods. The major attributes of the physical environment also demand that the formal research and extension system adapts its activities to reflect the overall limitations of the area, and respond appropriately to the specificity of conditions which farmers face.

3 Agricultural performances, economic diversification and household livelihood portfolios

As part of the reality which farmers must confront each day, the constant fluctuation and often marginal productivity of physical conditions found within the OHVN zone play a pivotal role in influencing farmers' individual behaviours, as well as the overall survival strategies pursued by rural households. The ability to prosper, and in many instances, simply to survive in such environments, requires that farmers possess both the capacity to formulate strategic management plans, as well as to take in stride and effectively handle diverse stresses, the full range of which can never be predicted or planned for a priori. Building upon the conceptualization by Richards of agriculture as a fluïd 'performance' (1989b; cf. Watts, 1983), this chapter explores some of the principal themes and characteristic adaptiveness of farmers' behaviour in these diverse and risk-prone environments. In shifting attention to an analysis of household production strategies, the notion of agricultural performances must be seen within the larger patterns of adaptive household economic diversification in agricultural, non-agricultural, and on- and off-farm activities. Large-scale variations in the constraints and opportunities faced by households in different areas of the OHVN zone lend themselves to characterization through the use of a 'household livelihood portfolio' perspective, capable of delineating the spatial trends in household diversification as one moves across the landscape.

These concepts—agricultural performances, household economic diversification, and livelihood portfolios—reflect central qualities of the local production systems, and serve as an important foundation for gaining a clearer understanding of farmers' individual and joint behaviours. In promoting the use of 'knowledge in action', a theme which runs throughout this book, it is suggested here that in order for external development agencies to achieve more significant and enduring impacts, they would be well advised to reflect upon and internalize these features of rural life in the design and implementation of their interventions.

Agriculture as performance

At the local level, the inherent temporal and spatial heterogeneity of natural resources is paralleled by an equally variable set of social resources—labour availability, production objectives, personal attributes of individual managers and the like. Within the context of annual crop production the convergence of these two sets of forces, the physical and social, is possibly best conceived as a dynamic performance (Richards, 1989b), or, more accurately, as a set of linked performances through which individuals respond to the evolving set of physical and social conditions within each year, and from one year to the next. The general nature of these performances is summarized by Moris (1991:27):

> African farmers respond to an unfolding scenario of changing opportunities and constraints (e.g. wet vs. dry years, early vs. late onset of the rains, pest outbreaks in a particular crop, unplanned labour shortages, and so forth)... [T]hey bring to their farming an accustomed 'script,' modified as the season

unfolds in a manner Richards (1989b) likens to a musical performance, or which Stewart (1986) describes as 'response farming'.

At a more intimate level these 'performances' embody the application of rural people's knowledge, skills and creativity in facing new and often unforeseen challenges. Yet, the question remains, what goes into these performances; how do farmers 'perform' agriculture?

At the outset of each production cycle, farmers have a general, yet fairly complete, plan of how they would like to allocate and manage their available physical and household resources (e.g. land, labour, equipment, seed stock, inputs). Farmers know which fields they would like to plant with what crops, using the seed stock they have saved, and (based upon field history) which fields they would like to manure, bring in or take out of fallow, and which fields will most likely need special care in handling such things as emerging weed problems. These plans are based upon both the accumulated 'wisdom' of traditional farming practices and elements of newly acquired management ideas or strategies. In any given year, however, constraints imposed by the status, quantity or quality of farmers' essential resources often cause even these initial plans to deviate significantly from what farmers know as 'best' practices. As the season unfolds, changes in the productive potential of available resources (e.g. human and/or draught animal health, equipment failure), coupled with unforeseeable delays or variations in the physical environment (e.g. rainfall, plant disease and pest outbreaks), require farmers to begin the process of further modifying and reformulating their initial assessments; the dynamic struggle of attempting to do what is 'best' within the realm of what is possible. This struggle is succinctly described by Richards in the case of intercropping, where the 'layout of different crops in a field is not a design but a result, a completed performance. What transpired in that performance and why can only be interpreted by reconstructing the sequence of events in time. Each mixture is a historical record of what happened to a specific farmer on a specific piece of land in a specific year' (1989b:40).

Viewed in this way, farmers' agricultural performances represent elements of both planned 'scripts', comprised largely of traditional practices, and provisional adaptations that farmers elect, or are required, to implement in response to unforeseen events. To expand Richards' (1989b) analogy between agriculture and music, these two features, tradition and adaptation, can be understood as representing the melodic themes and improvisational nature of farming in Sahelian West Africa.

It is important to stress that this type of adaptive improvisation does not imply (see Batterbury, 1993; 1996) that Sahelian farmers are incapable of conceiving or carrying out more strategic natural-resource management plans—the evidence to the contrary, within the OHVN and elsewhere, is overwhelming (e.g. Reij *et al.*, 1996; van den Breemer *et al.*, 1995; Warren *et al.*, 1995; among many others). Although the focus of a 'performance perspective' clearly centres on the adaptive in-season challenges that arise in annual food production, largely in response to the vicissitudes of the region's climate, there is nothing in such a perspective that under-values or denies farmers' capacity to plan for, or act in consideration of, longer-term resource management plans. The different time-frames of inter-annual natural resource planning (e.g. tree planting, construction of water and erosion control structures and soil fertility management), and the responses made by farmers to intra-annual production stresses, are neither mutually exclusive, nor, in terms of impact, easily separated aspects of agricultural management. As will be elaborated, many of the conditions to which farmers must respond in any given season are of their own making; case-studies continue to demonstrate farmers' capacity to both develop and adopt longer-term planning horizons as the need arises (e.g. Defoer *et al.*, 1996; Harris, 1996; Tiffen *et al.*, 1994). Nevertheless, farmers' ability to control, and even to significantly dampen the magnitude of some types of environmental variation, is limited. The immutable nature of the production environment will continue to require that farmers rely upon their adaptive management skills in order to get the most out of each season. As will become increasingly clear later in this and coming chapters (4 and 6), farmers both respond to the immediate, and plan for the future. For the present, however, attention will focus on better understanding the adaptive aspects of farmers' agricultural management behaviours.

Agricultural themes

Although a wide degree of variation exists between the individual practices used by different farmers, except under the most extreme conditions, farmers' general practices never wholly depart from a number of more general agricultural themes.[1] Certain elements of these basic themes can be seen in the data collected on cropping patterns. In the OHVN zone, as throughout much of the region, intercropping accounts on average for 70–80 per cent of the cultivated surface area, dominated by a dozen principal crops and their associations (DRSPR/OHV, 1992d). These major associations, reported in Table 3.1, illustrate some of the common elements of farmers' decision-making concerning intercrop mixtures—cereals paired with legumes, crops (and varieties) with different maturation periods and tolerances paired with each other—which are often created, in the words of one farmer, to increase the chance that a field would 'produce something'. The use of specific varieties and associations are further matched with specific soil types and moisture regimes; varieties of red-grained sorghum, tolerant of moist soil conditions, are planted in and along the annually inundated *bas fonds* areas and other 'wet' sites, while millet is planted in the drought-prone upland soils, and white-grained sorghum and maize, with higher nutrient needs, are planted in the free-draining inner and more fertile bush fields (e.g. Harlan and Pasquereau, 1969; Matlon, 1980; Stoop, 1986; 1987a; 1987b; Vierich and Stoop, 1990).

Other aspects of farmers' general 'themes' can also be seen in their preferred strategies for handling specific types of adversity. During years of low rainfall, and in the hopes of salvaging some kind of a harvest, farmers may thin certain crops and provide their principal cereal fields with an additional weeding to reduce the competition for scarce soil moisture, while other fields are, in effect, abandoned. In years when major replantings are necessary, farmers often abandon specific field types, such as the thin and drought-prone upland soils and the already inundated *bas fonds*, and concentrate their remaining resources on the more fertile inner and outer fields (see Matlon, 1980). During such years, or when the initial planting is delayed, farmers favour crops and specific varieties with shorter maturation periods. Table 3.2 illustrates some of the shifts in importance of specific food and market crops for men and women farmers during 'good' as opposed to 'bad' (although not disastrous) years.

Table 3.1 ***Major crops and cropping associations in the northern and southern OHVN zones*** **(Source: DRSPR/OHV, 1992a)**

Culture	Kanika (North) (ha.)			Déguéla (South) (ha.)		
	Type 1 Households*	Type 2 Households	Type 3 Households	Type 1 Households	Type 2 Households	Type 3 Households
Mil/Sorgho		0.10				
Mil/Niébé/Sorgho		4.43	2.65			
Mil/Niébé		2.00				
Sorgho		0.55	0.78	2.25		1.23
Sorgho/Niébé		2.49	1.45	1.00	2.42	0.90
Sorgho/Arachide		0.34	0.54	0.25	0.21	
Maïs		0.35	0.03	0.50	0.22	0.22
Arachide		0.15	0.23	0.22	0.18	0.30
Dah		0.10		0.06		
Vouandzou		0.25				
Mil		0.60	0.03	5.00	2.00	0.75
Mil/Niébé/Sorgho/Dah			3.00			
Maïs/Mil			0.15	3.00	1.67	3.00
Maïs/Mil/Niébé Riz			0.40	3.50	1.56	1.19
Coton/Niébé			0.05			
Sorgho/Dah			0.15			
Maïs/Sorgho				2.75	1.20	1.39
Coton				2.00	1.00	
Arachide/Dah				0.26	0.16	0.28
Maïs/Mil/Courge/Cala				4.00		
Fonio						0.25

* The three 'household types' refer to the categories of the household typology developed by the DRSPR/OHV (1993a). Type 1 households (representing 3.5 per cent of the households) are defined as those owning: at least 32 animal units (one cow equals 0.7 animal units one sheep or goat equals 0.12 units); six or more draught animals; 2 or more ploughs; and at least one additional piece of animal traction equipment. Type 2 households (20.5 per cent of the population) are defined as owning: between 5 and 31 animal units; 3–5 draught animals; two or fewer pieces of animal traction equipment. Type 3 households (76 per cent of the population) are defined as owning: four or fewer animal units; two or fewer draught animals; one or no ploughs.

Table 3.2. ***Change in importance of food crops in 'good' versus 'bad' years, in the northern and southern OHVN zones***

Kanika (North)				Samako (South)			
Men		Women		Men		Women	
Good	Bad	Good	Bad	Good	Bad	Good	Bad
Millet	Millet	Millet	Millet	Sorghum	Maize	Sorghum	Maize
(long)	(rapid)	Sorghum	(rapid)	Arachide	Sorghum (rapid)	Maize	Sorghum
Millet	Sorghum	Haricot	Sorghum	Millet/		Arachide	
(rapid)	(rapid)	Arachide	(rapid)	Maize		Rice	
Arachide							
Sorghum							

Agricultural improvisation

As important as these basic themes are to a general understanding of agriculture in diverse and risk-prone environments, it is still necessary to probe deeper into farmers' specific behaviours in order to appreciate the more improvisational aspects of local farm management. For instance, detailed studies on farmers' intercropping practices within the region have found as many as 23 different crops involved in 230 to over 300 intercropping associations (e.g. Norman *et al.*, 1979; World Bank, 1979). Many of these associations are planted in only a portion of an entire field, often sown at a later date, by a different 'manager,' as time, conditions and resources permit. In one such example, a 'farmer' in the northern OHVN had planted *dah* (*Hibiscus spp.*) in four different ways in a small corner of the household's main sorghum field: as a border crop; as alternate plants within a single row; in alternate rows; and in the furrow between rows. The data collected on intercropping, however, typically treat one sorghum-*dah*, or millet-cowpea, intercropping association as being the same as the next. The use of specific varieties, planting densities and variations in the temporal and spatial arrangements of the different associations are not accounted for, yet are critical for understanding the complete story of how and why a particular association was established, as well as some of the agro-ecological interactions taking place (e.g. Carsky *et al.*, 1994; Ntare and Williams, 1992).

Not only can statistical generalizations obscure important aspects of farmers' behaviour, but, as cautioned by Richards (1989b), a priori assumptions about farmer behaviour must not be confused with the apparent logic of farmers' actions. The result of a completed performance is often a by-product, or artifact, of a specific technology or set of decisions made in response to a particular set of circumstances, rather than evidence of a concerted effort to plan for those results on the part of a farmer. To illustrate the point, one common labour-saving planting practice used by farmers is to mix millet and/or sorghum and cowpea seeds to a desired proportion in a small gourd bowl for planting. 'Pinches' of this mixture are then rapidly sown with a short-handled planting hoe, *daba*, in holes spaced according to the stride used by the individual performing the operation. While the overall stand density and general populations of the crops sown can be controlled, the exact intercropping pattern that emerges is established largely by chance. In this instance, the resulting intercrop mix is more a product of the need to get a field sown quickly, than an attempt to maximize 'a general theory of inter-species ecological complementarity' (Richards, 1989b:40).[2]

This is not to say that farmers do not understand and often utilize such types of 'complementaries.' An alternative, more labour-intensive, practice used by farmers in establishing millet-cowpea associations is to delay the planting of cowpeas in cereal fields in order to minimize moisture competition with the primary cereal crop during early growth, as well as to reduce the risk of an early season drought harming the cowpea crop, which is more susceptible during its early development. The level of control achieved by this practice does allow farmers to implement a 'general theory of inter-species complementarity', yet it comes at a cost in terms of labour and time, which farmers are not always willing or able to meet; the conditions of a particular season must, therefore, allow a particular strategy to be pursued.

A large measure of the individual agricultural performances enacted by farmers depends upon their perceptions, assessment and skills in responding to real-life conditions as they unfold. Hypothetical discussions, for example, concerning farmers' general responses to early or mid-season drought may provide a clear indication of what farmers would choose to do, all other variables remaining constant, but they should not be equated with what a particular farmer may be capable of doing, or elect to do, in any given year. Even farmers' descriptions of how they responded to past stresses is not always instructive of how they will respond to similar stresses in the future. Such discussions are 'out of time' (Richard, 1989b:40), devoid of the intervening 'in time' circumstances of Sahelian life that constantly force farmers to alter their behaviour.

Observations of farmers' 'in-time' behaviour under real-life conditions reveal the dynamic process of adaptation that farmers engage in when responding to multiple, and often countervailing, influences. Watts (1983), for example, describes a number of different types of adjustments in field use, plant spacing, thinning, over-seeding and crop and varietal selection that individual farmers used in response to particular stresses. In their studies, Mortimore (1989) and Toulmin (1991) provide details on the broader, and very different, inter-annual investment strategies followed by households with varying asset levels. Such examples depict the underlying improvisational nature of agricultural performances—aspects that portray agriculture as the product of a series of decisions, made 'on the run,' influenced by a farmer's knowledge, creativity and her or his access to resources, rather than as the implementation of a unitary plan or a series of fixed contingencies.

Understanding the 'performance of agriculture' as a series of inter-linked acts, or suite of movements, is particularly important in that it helps to bridge the seemingly disparate qualities of improvisational and planned management. At any point in the cropping cycle, many of the conditions to which farmers must respond are at least partially a construct of their previous performances, both within a single season, as well as from previous years. Decisions regarding where, when, how and what crops and varieties to plant follow (and are made partially in response to) the conditions resulting from previous decisions concerning the negotiated access and use of land and labour, the timing and method of land preparation, the build-up of weed populations, change in soil fertility levels and existing food stores and seed stocks. Once completed, the act of planting is followed by equally important sets of decisions over the timing and method of weeding (including labour negotiations and payments), thinning, transplanting, any over-seeding and, finally, harvesting. Viewed from a 'systems' perspective, farming is a process where the ending point of one set of decisions, and season, help define the starting point for the next. However, it is also a process over which farmers have limited control. In risk-prone environments, farmers' ability to create conditions (e.g. fertile fields) is equalled, if not surpassed, by the creation of conditions (e.g. insufficient, or ill-timed rainfall), to which they must respond. A full understanding of the performance nature of agriculture must, therefore, be based upon both a knowledge of farmers' desire and struggle to build a better future, and the necessity of meeting the many challenges of the present.

Household economic performances

Many of the characteristics of farmers' agricultural performances are also visible in the overall economic strategies pursued by rural households. Depending upon the conditions of each season, the availability of resources, and the particular household's preferences and needs, investments are made in a wide range of economic activities that not only provide security, but also serve to more fully exploit the perceived opportunities. The seasonality of agriculture, combined with the high level of insecurity of returns within seasons and opportunities outside of the sector, lead to households being best characterized as 'multi-enterprise production units... far from purely agricultural...[where]... household members are typically part-time farmers and are simultaneously engaged in a variety of non-farm production activities' (Hunt, 1991:49). Farmers in one study conducted in the OHVN identified over 120 different sources of income (Sundberg, 1989). Other studies in the area, as well as elsewhere in the region, have shown that non-agricultural activities provide from 40 to nearly 70 per cent of household revenues (DRSPR/OHV, 1992c; Matlon, 1977; Reardon *et al.*, 1992; 1994). The importance of this point is further emphasized by the fact that a majority of households in the central OHVN zone are not self-sufficient in basic grains; roughly 20 per cent of the net grain consumed is purchased, with a greater share being purchased during years of poor production (see Steffen, 1992).

Of those factors influential in shaping the extent and forms of household economic diversification, household wealth and labour availability are possibly the most important (e.g. Grosz-Ngaté, 1986; Koenig, 1986a; Lewis, 1979). Poorer households, which in most cases are those with the fewest members, diversify through necessity; they simply are unable to meet all their needs through agriculture alone. The overall extent and nature of the poorer households' diversification, however, is limited by their generally small labour force and low level of resources, which can act as a barrier preventing them from entering certain labour markets (e.g. Reardon, 1997). Wealthier households, on the other hand, tend to diversify based on ability; sufficient labour and accumulated capital allow these households to release resources from agriculture for investment in other activities (Reardon *et al.*, 1992). From seasonal labour migration (e.g. Mazur, 1984; Reardon *et al.*, 1992; Van Westen and Klute, 1986) to the gathering of economically and nutritionally important resources from common property reserves (e.g. Bergeret and Ribot, 1990; Falconer, 1990; cf. Scoones, *et al.*, 1992), household decision-makers choose from a broad spectrum of alternative activities in attempting to maximize their opportunities, maintain social solidarity, and reduce economic uncertainty and risk.

Just as with the flexibility and adaptiveness exhibited by individual agricultural performances, the allocation of resources in various economic activities at the household level is not static. Year-to-year changes in conditions within and outside of the household, tempered by the skills and preferences of individual household members, ensure a constant evolution in investment patterns. Understanding the fundamental significance, breadth and fluidity of household diversification will be an important starting point for any organization interested in improving the economic welfare of rural households in productively marginal and risk-prone environments.

The social organization of household production systems

Basic food security, as well as most decisions on economic diversification, depend upon and evolve out of the social organization of the household—i.e. who has access to, can mobilize, or controls what resources. Based upon field observations and several detailed studies on the social organization of land, labour and other resources in Bambara and Malinké society (e.g. Becker, 1990; Grosz-Ngaté, 1986; Jones, 1976; Koenig, 1986a; Lewis, 1979; Toulmin, 1991; 1992; Vierich, 1986), a number of generalizations can be made regarding the organization of household production systems in the OHVN zone.[3]

At the village level, social organization revolves around a male-dominated hierarchy of lineage and age. Typically the eldest member of the eldest generation (age-set), generally from one of the village's founding families, is the village chief. Individual family units are grouped into compounds, or *du*, where the various dwellings, pens and appendages form a self-contained unit surrounding a central courtyard(s). Depending on the village, individual compounds can be linked together, organized into neighbourhoods or wards established along ethnic or lineage lines, or separated into a loose federation of individual concessions spreading over a square kilometre or more. Some villages that have experienced significant growth have splintered, frequently as a result of resource pressure (e.g. land, water) and/or social tensions, into a number of separate hamlets located at some distance from each other.

Within the village, the most important unit of social organization is the extended family household. Presided over by the eldest male, households are commonly referred to as the *gwa* (meaning literally 'the hearth')– 'the unit that farms a common field and eats most, if not all, of its meals from a common granary' (Toulmin, 1991:117). The formation of polygynous households is common throughout the OHVN zone. Family groups with more than one *gwa* may be showing signs of imminent breakup (Lewis, 1979). On the whole, demographic data for central Mali show that households are comprised of a bi-modal population of larger complex households with as many as 23 to 30 members (72–84 per cent), and smaller, generally newly-formed and economically insecure, households of 7 to 10 individuals (e.g. Becker, 1990; Toulmin, 1986; 1991;). Recent trends appear to be towards smaller household units, which Becker (1990) interprets as a form of organizational innovation in response to an increasingly market-oriented rural economy (cf. Jones, 1976).

Under the management of the eldest active male, household members hold first allegiance to the cultivation of their main subsistence cereal crops. Tasks are generally gender-based, although the exact division of labour can vary considerably by household, depending to a large degree upon the interpersonal relations and attitudes of the household members. In general, men perform the heavy tasks of clearing fields and use of animal traction, with women doing much of the planting, weeding, and harvesting. Other activities, such as winnowing grain and all domestic chores, are exclusively a woman's task and are co-ordinated under the leadership of the eldest active woman in the household. By the age of 10, children regularly accompany their parents into the fields and have begun learning the tasks associated with their separate gender roles in earnest (e.g. Cross and Barker, 1991). By the age of 12 to 14, most children are active workers in the communal fields (Grosz-Ngaté, 1986).

Beyond the contribution of labour in cultivating the main household fields, adults may also cultivate one or more joint or personal fields subject to land availability, the adequacy of household labour, group norms and individual negotiations. The granting of land and labour concessions is one way in which household heads attempt to maintain harmony and solidarity among family members; this is critical to the successful mobilization of household labour, which is the major challenge among larger households. As Lewis (1979) notes, however, personal fields detract from the yield performance of communal fields, and within Bambara communities a household whose members operate a number of individual fields may be showing signs of internal discord.

Once secured, the location of personal fields, types of crops grown and end use of the produce differ significantly by gender. For men, individual fields are generally located in the more accessible and fertile bush fields, and are planted to a cash crop from which some minor gifts may be granted to other family members. Depending on their age and the status of local land availability, women may cultivate a small upland field of peanuts and/or millet intercropped with other species, or rice in the *bas fonds* if such an area exists. Women are generally able to secure access to individual fields once they have become the senior wife in control of the hearth and their labour is no longer needed in the communal fields (often synonymous with menopause, personal communication Arnoldi; cf. Grosz-Ngaté, 1986).[4] In some locations, younger women in the household may jointly manage a separate field (cf. Grigsby, 1989). In others, as with the case in the southern OHVN where severe land shortages exist, women may be able to secure *usufruct* rights to their husbands' fields only during the peanut rotation in the cropping cycle. At times, even young girls can secure a small plot of valuable rice land for themselves by doing favours for an older woman, such as bringing food out to the fields or helping with the weeding. Women typically use their personal fields to generate cash, cement social ties, assist their children and supplement the household cuisine (Lewis, 1979; Grosz-Ngaté, 1989). In addition to field crops, most married women, regardless of age, manage a small garden plot, which is an important source of personal income and condiments for the household.

Aside from the intra-household demands for agricultural labour, both men and women take part in a number of other labour-sharing arrangements and remunerative work groups outside of the household. Depending on the solidarity of the village, this may include work in the fields and other activities of the village-wide social organizations (*tons*), which are divided by gender and age-sets (for details, see Grigsby, 1989; Lewis, 1979; Leynaud, 1966; Leynaud and Cisse, 1978), ward labour, or more *ad hoc* labour arrangements, such as house or granary construction for men, or harvesting and winnowing grain for women. Although there may be some outright granting of assistance to poorer households, most loans of labour, equipment or animals involve the remuneration of cash, grain or labour in-kind. Depending upon need, households rely upon a number of different types of extra-household labour arrangements during the course of any one year, or in some cases, over periods of several years. Although the exact nature of the arrangements (individual or group), type (hand labour, animal traction or tractor), duration (day, piece-work, season, or longer), and form of payment (cash, in-kind, or produce), differ widely, the ability to organize and mobilize these alternative forms of labour plays a vital role in determining the current and future welfare of most rural households (for more detailed discussions see Grosz-Ngaté, 1986; Lewis, 1979; and Toulmin, 1991; cf. Berry, 1989).

The diversification of household economic activities

Agricultural surplus and cash crops To generate income, farmers from all areas of the OHVN zone rely upon the sale of surplus food crops and cultivation of one or more cash crops. While the sale of surplus grain is generally dependent upon the growing conditions of that year, the decision to engage in cash-crop production reflects individual preferences, household capacities, and factors including agro-ecological limitations, market proximity, access to transportation and level of institutional support. The staples most commonly sold are millet, sorghum, fonio (*Digitaria exilis*) and maize, while common cash crops include cotton, peanuts, tobacco, garden vegetables (e.g. tomatoes, onions, okra, eggplant, pepper, and many others), fruits (e.g. mangoes, citrus, bananas, water-melons), livestock and animal by-products (e.g. eggs, milk and, at times, manure). Separate from the sale of grain for consumption, farmers also sell limited amounts of seedstock.

In some of the more remote villages visited, farmers expressed a reluctance to engage in grain sales that resulted in grain leaving the village (cf. Lewis, 1979). In one village, farmers involved in a seed multiplication programme had sold very little of the improved seed they had produced, preferring instead to store most of it in the project warehouse as a form of village food security. Other communities have independently, or through the assistance of an NGO, built their own communal grain stores into which each household contributes grain. These communal granaries are then used as a sort of village bank, not just for food emergencies, but for economic needs as well, such as to cover the expense of medicines. Such general sources of credit not only relieve some of the pressure on individual households, in terms of lending, but also change the nature of indebtedness from personal to communal, where the burden is shared.

On-farm agricultural activities In addition to the direct sales of agricultural produce (crops, seedstock for planting, animals and by-products), households also rely heavily upon earnings from a number of on-farm value-added processing activities. Farmers identified over a dozen major income-generating activities involving the transformation of produce and by-products from their cropping systems (see Table 3.3).

Many of these activities involve the processing of indigenous 'secondary' crops and tree products that are an integral part of local production systems. In areas with extensive '*karité* parks', upwards of 90 per cent of the households engage in processing *karité*, with as much as 60 per cent of individual women's annual income coming from the sale of products made from *karité* oil alone. Local economies in many areas are driven by sales of *karité* products to the extent that, in years with low nut production, local market activity is noticeably depressed (Grigsby, 1989).[5]

Table 3.3 ***Major on-farm value-added processing activities***

Garden vegetables (dried; fermented)	Tamarind (dried; beverage)	Peanut and Cowpea hay (forage)
Karité (*Vitellaria paradoxa*) (butter; soap)	*Dah* and Sisal fibre (cord)	Baobab (*Adansonia digitata*) (dried leaves and fruit)
Néré (*Parkia biglobosa*) (powder; soumbala)	*Henné* (*Lawsonia inermis*) (dye)	Skins (dried and tanned)
Dah (dried; fermented)	Sorghum and Millet stalks (mat construction)	Prepared foods and cheese

On-farm non-agricultural activities On-farm, non-agricultural activities include the gathering of fuelwood, cutting fodder, collection of wild and cultured honey, charcoal production, pottery making, petty commerce and numerous artisan trades (see Table 3.4). These activities require varying degrees of skill and/or the use of specialized equipment. In the larger towns of the zone, sufficient consumer demand and market integration allows individuals to specialize in certain enterprises as bona fide tradepersons (e.g., merchants, tailors, bakers, butchers, jewellers). In most rural areas, however, these activities fall far short of full-time vocations, and in no way should be confused with the occupations associated with specific ethnic groups and castes, such as herding, commercial fishing and blacksmithing, or as Lewis notes (1979), even with such specialized activities as mending calabash, which is performed by women of the woodworker caste.

Many of these non-agricultural activities involve the exploitation of local common property, or open access, resources.[6] In the case of different forest resources, villagers throughout the zone reported a range of ownership patterns. For example, the majority of 'wild' fruits and other tree products are treated simply as open access resources, available to anyone who chooses to invest the time collecting them. Specific species, on the other hand, such as the *karité* and other important species, often have different degrees of control applied to access; trees on fallowed lands are treated as a common property resource, while those in active fields may become the temporary 'property' of the household managing the field. Other species, such as the baobab, are often privately owned, either through inheritance, planting, or on a less-permanent basis, established by pruning their branches to increase productivity (Lewis, 1979).

In general, villagers reported that these local, non-agricultural activities involving open access and common property resources are of greatest importance to the incomes of women, the elderly and economically insecure households. This view is reinforced by several studies that identified the sale of 'wild' fruits as the most important income source for women in the OHVN (BECIS, 1991;

Table 3.4 ***Major sources of on-farm non-agricultural income***

Bee-keeping	Jewellery making	Wood cutting	Bricklaying
Baking	Hunting/Fishing	Petty commerce	Cloth dying
Tailoring	Pottery	Basket weaving	Charcoal production
Traditional medicines	Carpentry	Cloth making	*Secco* construction
Well-digging	Midwifery	Yarn spinning	Repairmen

DRSPR/OHV, 1993c; Luery, 1989; Sundberg, 1989), and the fact that the majority of people selling 'bush' products in the markets are women (cf. Guinko and Pasgo, 1992).

Off-farm employment For most households in the zone, the economic contributions from off-farm employment, both local agricultural and non-agricultural, as well as remittances from those who have gone on labour migration (*exode rural*), are another important source of income to the overall financial security of the household. Off-farm employment is pursued in all seasons, although it is most pronounced in the dry season and during years of poor rainfall, when agricultural harvests as well as the production of many of the secondary products used in other income-generating activities are diminished. During periods of severe drought, poorer households are among the first to shift attention away from their own fields and begin selling their labour to others in an attempt to keep themselves fed (see Mortimore 1989; Watts, 1983). Such conditions can result in household members being 'pushed' off the farm, to help reduce the drain on limited food supplies.

In addition to covering shortfalls in household food production, investments in off-farm employment also serve to take advantage of surplus labour and capitalize on unique money-making opportunities (e.g., Lewis, 1979; Reardon *et al.*, 1992). During the growing season the most common arrangements are for day labour (by a group or individual), piece work (e.g. ploughing of a field), and occasionally seasonal, when an individual is 'borrowed', or in the case of hardship, 'accepted', into the household for an entire growing period or longer. During the dry season, off-farm employment beyond the village is generally sought for periods lasting from three to nine months. Multi-year sojourns are also common (see DRSPR/OHV, 1991a), and involve both agricultural and non-agricultural labour within the region. Common destinations for off-farm employment are nearby urban centres, the capital city, Bamako, and several key international destinations. While in the past international opportunities were commonly sought, farmers reported that in recent years international travel has become less important as the domestic labour markets have improved, and international markets have become less certain (due in part to the effects of regional conflicts, as well as national policies aimed specifically at stemming the flow of immigrant labour, e.g. in Côte d'Ivoire). In some rural areas, the increasing awareness and concern about AIDS has resulted in a social stigma being attached to those who have migrated to neighbouring countries, making it difficult for men to find wives upon their return (pers. com. Gnägi). Other social problems can befall individuals who have stayed away 'too long' (Lewis, 1981).

Due to its impact on local agriculture, one of the most widely discussed and criticized (by the extension service) off-farm employment activities in the OHVN zone is the operation of gold placer mines. For many individuals from villages along both sides of the Niger River in the southern part of the zone, the operation of placer mines is one of their most important dry-season activities. Its importance varies by location, with as many as 95 per cent of the households in some villages reporting economic benefits (DRSPR/OHV, 1991a). Those involved in gold digging also vary by village. In one village, 75 per cent of the young men (and few women) were involved in digging, while in another village, gold digging occupied 90 per cent of the women (DRSPR/OHV, 1991a). On a daily wage basis, women who work in the gold fields can earn wages comparable to those obtained from group field labour. Yet in gold digging, women can individually keep all of their earned wages, unlike in group labour, where wages often go into a group *caisse* (Luery, 1989). The lure of easy profits keeps many in the gold fields until the onset of the growing season, preventing them from the timely preparation of their fields and thus affecting their yields.

Household livelihood portfolios

The general economic opportunities and constraints faced by households in different areas of the OHVN zone are shaped by the combined influences of variances in the physical environment, proximity to markets, access to reliable and affordable means of transportation, and institutional histories, among other factors. One way of representing the resulting patterns of household diversification is through an examination of the constellation of agricultural and non-agricultural, on- and off-farm economic activities in which households invest resources in meeting their consumption and economic needs (cf. Gupta, 1995; McCorkle and Kamité, 1986; Reardon *et al.*, 1992). The concept of household livelihood portfolios is introduced here as a descriptive tool for helping to characterize household responses to the broad range of physical, social, and economic conditions found within the OHVN zone:

> The idea of a household economic portfolio draws upon a conceptualization of household decision-makers as investors who allocate their scarce human, financial and physical resources according to perceived short- and long-term, low and high risks. Like those who play the stock market in industrialized societies, rural producers constantly scan the investment horizon to identify the best ways of allocating their resources to protect and improve their standard of living (Bingen *et al.*, 1993:3).

Similar to farming-systems research recommendation domains (e.g. CIMMYT, 1980), a portfolio perspective can be used to identify the geographic range of distinct cropping systems (see Table 3.5) that correspond to the agro-ecological, as well as the economic and institutional, conditions of different areas. Yet unlike recommendation domains, which limit their focus to agricultural activities alone, the portfolio perspective helps to locate the relative importance of these agricultural activities within the larger context of a household's overall economic performance. For development organizations interested in improving the economic welfare of rural households in marginal environments, adopting such a

Table 3.5 *The transition in importance (per cent of agricultural land) of cereal crops by secteur* (Data from OHVN, 1992a)

Secteurs (north to south)	Millet (percentage of ha.)	Sorghum (percentage of ha.)	Maize (percentage of ha.)	Rice (percentage of ha.)
Boron	76	24	< 1	–0–
Banamba	67	33	< 1	< 1
Sirakorola	64	35	1	< 1
Koulikoro	50	46	3	1
Kati	16	53	29	2
Gouani	25	60	14	1
Dangassa	24	55	16	5
Oueléssébougou	30	58	8	4
Bancoumana	16	54	23	7
Kangaba	11	36	33	20

framework for understanding and monitoring the breadth and fluidity of household diversification will be an important point of departure.

Building up from the investment patterns of individual households and communities, the distinct characteristics of portfolio areas do not emerge until a sufficient level of aggregation has been reached; well above the individual household and village level, yet below the sub-national generalizations based upon agro-ecological criteria alone. Because portfolio areas reflect a composite of household opportunities and constraints, each of which has its own geographic limit, the boundaries between different portfolio areas are, by necessity, blurred. The similarity of conditions in each area begins to break down into broad transition zones, resembling, in many ways, the 'patches' of landscape ecology, where increasingly frequent pockets of households begin to have access to and are able to take advantage of new opportunities.

As suggested earlier, the exact composition of activities undertaken by any individual household is most notably influenced by wealth and the size of its labour force. As a result, inter-household differences within portfolio areas may actually exceed the differences in profiles between portfolio areas. However, while inter-household differences (i.e. wealthy and poor households) are prevalent throughout the zone, the specific constraints and opportunities facing households in different locations, regardless of their size or level of wealth, have geographic limits (e.g. cotton production, the marketing of perishable fruits and vegetables, gold mining, proximity to certain labour markets, and so on). In addition, household wealth levels are not constant. The accumulated effects of drought, loss of animals, sickness, family divisions and strings of both 'good' and 'bad' luck can cause the fortunes of individual households, and entire communities, to be transformed literally overnight. Over the course of several years, households may view the world from a number of different perspectives, from that of the wealthy to the impoverished. Thus in order to operationalize the portfolio concept for programming purposes, it will be important to account for (and monitor) the differentiation between household wealth levels within locations, such temporally bound detail is not however as vital for the present purpose of providing a general description of the overall possibility frontiers of major opportunities and constraints facing households in different areas of the OHVN zone.

The Far North Portfolio Throughout much of the Boron and the northern portions of Banamba *secteurs*, low and highly variable annual rainfall (<600mm) strongly influences the range of household agricultural and other investment opportunities (see Figure 3.1).[7] Common to most of West Africa, farmers in this area and the rest of the OHVN zone use a system of 'inner' fields that are under continual use, fertilized with animal manure, household wastes and occasionally compost, and 'outer' fields, managed under a fallow system, to maintain soil fertility and help spread their production risks (cf. Prudencio, 1993). The cultivation of short- and long-season local varieties of millet, and to a lesser extent sorghum, is a central feature of the Far North Portfolio. Farmers commonly plant several varieties in each of their fields to improve the probability that they will be able to harvest something for their efforts. Cowpeas (*Vigna unguiculata*) are often intercropped with cereals, although when grown in these associations the cowpeas are generally destined for use as fodder and, as one farmer explained, are used not because the cowpeas are known to benefit the production

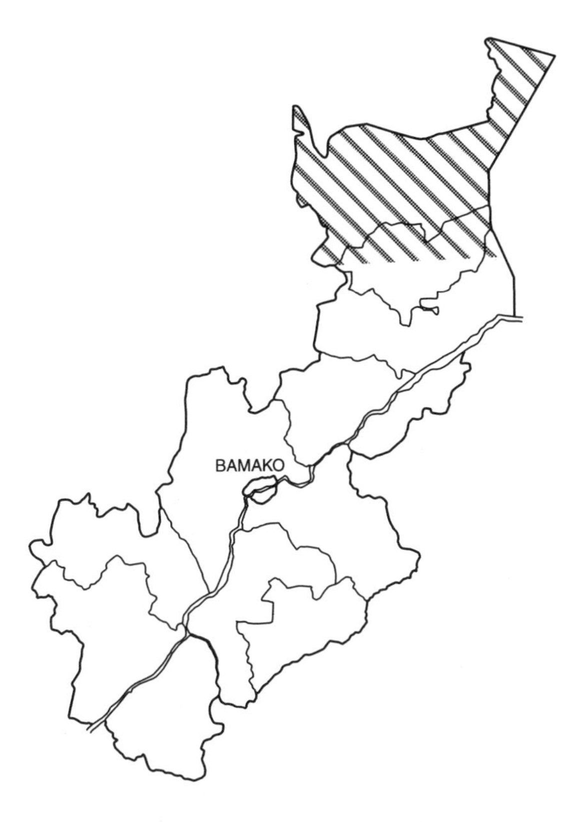

Figure 3.1 *The Far North Portfolio area*

of the millet, but because it is a way of getting something extra out of the field without hurting the production of the principal cereal crop. Other crops, such as *dah*, which are commonly intercropped in cereal fields further to the south, are planted separately in the Far North because of the noted negative impact on millet, probably due to moisture competition. Households use separate fields to optimize the production of cowpeas (for consumption) and other crops, such as fonio, and small parcels of *henné* used in making dye.

Farmers still rely upon peanuts, often intercropped with Bambara groundnut (*Vigna subterranea*) and incidental millet plantings, as their main cash crop. The importance of peanut cultivation, however, has declined since the dissolution of the Opération Arachide et Cultures Vivières (OACV) in 1980, the governmental development organization that had promoted cash-crop peanut production in the area. As has been attempted in previous periods of development effort, the extension service is currently attempting to promote the production of sesame as an alternative cash crop for farmers in this northern area. Limitations placed on the distribution of planting material and marketing guarantees, however, have prevented the widespread cultivation of sesame.

In their cropping systems, farmers utilize a mixture of traditional, adapted, and 'improved' cultural practices. Farmers estimate that 70 per cent of land in this area is cultivated using animal traction; the rest is cultivated by hand labour. Although bullocks are preferred, the use of horses as draught animals is common, and dominant in some areas. Donkeys are only rarely used as draught animals, in contrast to farmers in other parts of Mali and neighbouring Sahelian countries who use them regularly. Other than the plough, the use of other animal-powered implements is low. Cultivation of most fields begins after the first rains, although some farmers plough and plant before the rains begin, a practice that was more common in the past under a somewhat higher and more certain rainfall regime. Most fields are prepared using ridges, which help conserve soil moisture and deter erosion, although with regards to erosion farmers in this area are much less concerned about soil loss than with keeping young plants down slope from being buried. The use of ridges requires fields to be hand planted, because the available mechanical seeders cannot be effectively operated along ridges. As noted among farmers elsewhere in the OHVN zone (e.g. Moseley, 1993), and region (e.g. Gubbels, 1992), the use of traditional soil conservation techniques such as rock and vegetative barriers in this area has fallen into disuse, possibly due to the influences of ploughing.

In association with their field crops, farmers retain and manage a number of important income-generating tree species. Although nothing like the dense 'agro-forestry parklands' found further to the south, isolates of baobab, *néré*, *duguru* (*Cordyla pinnata*), and desert date (*Balanites aegyptiaca*), among other species, are commonly found in farmers' fields. The contributions of *Acacia albida* in soil-fertility management is widely recognized, but not practised in all locations, due in part to the belief that the trees, which are leafless during the growing season, encourage increased bird predation by offering ideal resting places. The major droughts and change in rainfall patterns over the past several decades have led to extensive mortalities and the near complete elimination of several species from fields in many localities.

The investment in livestock plays an extremely important role in household financial security in this area, where animals serve as 'walking banks.' Although some smallstock (sheep, goats), young cattle and draught animals are kept in the village, most other livestock are placed under the care of Peul herders and taken into the bush. This relationship between the Peul and farmers, common throughout the OHVN zone, is based upon the complementary management specializations of herders and cultivators, and allows agriculturalist households to diversify their investments into both crop and livestock production, and provides herders with income and access to agricultural land. As detailed by Toulmin (1992), herders and agriculturalists also enter into reciprocal relationships in the exchange of crop stubble and access to water for manure. In the past, these relationships often formed an enduring part of the local social structure, and were passed on from one generation to the next.

In addition to the cash earnings from peanuts and surplus millet production, farm households, and especially women, rely upon a range of less well-known agricultural products as important sources of revenue. Some of these include the sale of *A. albida* seed pods as fodder, *henné* dye, the limited collection of wild and semi-cultivated fruits (including those from *karité*, baobab, tamarind, *duguru*, among others), gum arabic (*Acacia senegal*) (gum from *Acacia seyal* is also collected), fresh milk, meat and hides, and various leaves used in sauces. Households sell some of these products fresh, but process most others using traditional drying and fermentation techniques. Largely through the assistance of NGOs, women in this area have become increasingly involved in small-scale dry-season market gardens. However, market proximity, low effective demand and water shortages limit current opportunities. As found throughout the OHVN zone, households in the Far North also selectively pursue various other types of small-scale, skilled and unskilled income-generating activities, including wood gathering, carpentry, blacksmithing, butchery, basket-making, petty commerce, cloth making and dyeing, and well-digging, amongst others. Income generation and expenditure studies reveal that, unlike areas further to the south, these sorts of trades provide men in this area with their most frequent, if not most important, sources of income (Sundberg, 1989).

The Near North Portfolio Immediately to the south, including the southern portions of Banamba *secteur*, all of Sirakorola *secteur*, and the northern parts of Koulikoro *secteur*, is the Near North Portfolio area (see Figure 3.2). Due to higher rainfall levels, averaging between 600mm and 800mm per year, and greater marketing opportunities, this area supports more diverse cropping systems and associated economic activities. Local varieties of millet and, increasingly, sorghum remain the principal cereal crops which, as in the Far North, are often intercropped with an incidental cowpea crop. Farmers in this area have tried the improved cereal varieties, introduced through the FAO Seed Multiplication Programme and OHVN extension efforts, but report that the yield performance and taste do not compare with local varieties, and thus have not been widely adopted. The cultivation of fonio is common. Fonio is generally planted as the last crop in a field before it is fallowed, when the soil has become too

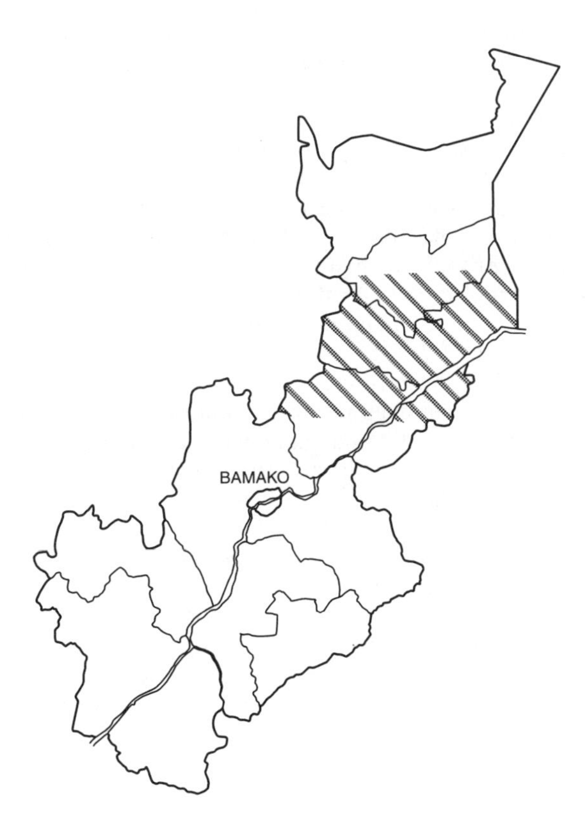

Figure 3.2 *The Near North Portfolio area*

impoverished to produce another millet or sorghum crop. Under these conditions it produces relatively high yields, commands a high market price throughout the year and is cultivated for both home consumption on important occasions and for sale. As with farmers in the Far North, those in the Near North are also interested in the economic possibilities offered by sesame production but, as yet, access to seed stock and market guarantees, strictly controlled by the extension service, have been limited.

The higher rainfall allows farm families to complement their staple cereal and legume production with the garden-type production of several vegetable and leaf crops, along with an 'insurance' crop of manioc. Farmers also use their gardens as nurseries for *henné*. The young plants are then transplanted into the main fields where they may remain for up to ten years. *Pourghère* (*Jatropha curcas*) and other unidentified species are used as living fences to protect these gardens and along field boundaries. In addition to its utility in helping to control livestock, *pourghère* is valued for its medicinal properties and is widely used in traditional soap production.

Significant soil and climatic variations in this area create a variety of localized production environments capable of supporting a range of unique investment opportunities, including the limited cultivation of maize, cotton, and even small quantities of rice. Small plots of specialty crops, such as sugar cane, sweet potatoes and water-melon, are also common. Locally grown mangoes are found in the Banamba market, as are other high-value tree products from the increasingly dense agroforestry associations of *néré*, baobab, *dugura*, desert dates, *karité*, tamarind (*Tamarindus indica*) and 'wild' fruits such as *nsaban* (*Saba senegalensis*). In addition, a growing number of women are involved in producing dry-season vegetables for sale in local markets and for home consumption.

Additional income-generation opportunities arise from the value-added processing of numerous agricultural products and by-products: *karité* butter; the preparation of soumbala from *néré* seeds; *henné* dye; mats made from the stalks of long-season sorghum and millet; and cord made from *dah* fibre. Natural resource extraction, through fuelwood cutting, honey collection, traditional medicines and other non-timber products gathered from the dry-land forests, offer important sources of revenues to certain households, as do the occasional sales of manure for use as fertilizer.

Many households rely upon earnings from dry season employment elsewhere in the country. Short-term labour migration is common, especially during years of poor rainfall. Under the most extreme circumstances children are sent off to live with kin located in other parts of the country, for a year or longer, where they contribute labour for their keep.

The Bamako Central Portfolio In the areas centred around Bamako, covering portions of eastern Kati, southern Koulikoro, northern Bancoumana and Dangassa, and parts of western Gouani *secteurs*, the increasingly favourable and secure annual rainfall of this area (800–1000mm) permits the production of a variety of fruit and vegetable crops in addition to staple cereals and legumes. The major factor influencing this area, however, is the easy access to the Bamako markets, which encourages a wide range of agricultural and off-farm activities. The impact can be seen to affect communities as far away as Koulikoro to the north and Oueléssébougou to the south-east (cf. Akasaka, 1973), where market gardens supply large quantities of tomatoes, peppers, okra, eggplants and other vegetables to buyers from Bamako (see Figure 3.3).[8] Additional niche markets

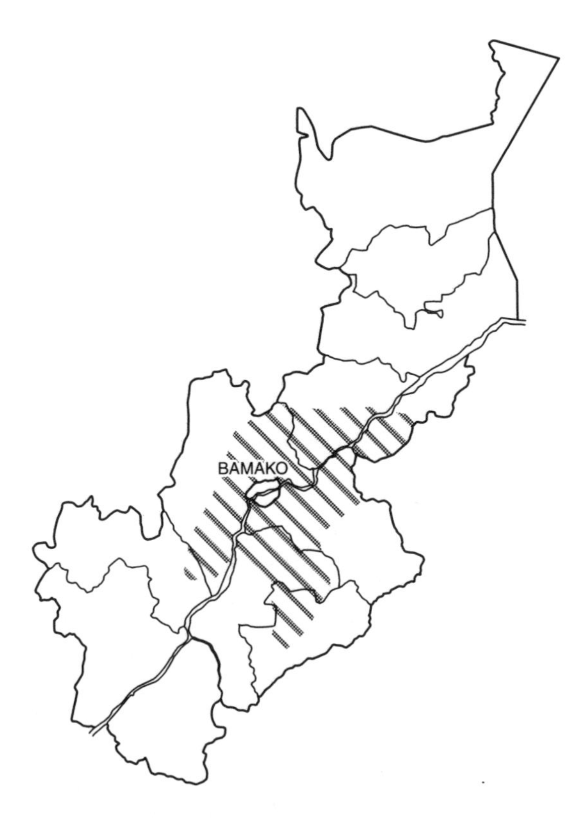

Figure 3.3 *Bamako Central Portfolio area*

include fodder production for urban livestock, market-oriented animal production (meat, eggs and milk), honey, 'wild' fruits and even chew sticks (a toothbrush substitute). The easily accessed urban businesses offer significant off-farm employment opportunities.

The Bamako Portfolio area also includes several special classes of producers that are not found elsewhere in the OHVN zone. These include the very small, but intensively managed, urban and peri-urban market gardens found throughout Bamako proper, which supply a significant share of the fresh vegetables consumed in the capital. Because of the highly specific nature of these urban garden systems, plus special land tenure issues, they constitute a definite sub-set of the Bamako portfolio (see Diarra, 1975; Swindell, 1988). A second special class of producers are the numerous absentee landlords, government officers and traders from Bamako, who constitute the major commercial investors in crop and animal production in this area. Their small to medium-scale plantations and livestock operations are widespread throughout the peri-urban area and along the major roads leading in and out of the capital. The early expansion of these plantations were among Malian civil servants who were given long-term leases to the land by the government (Jones, 1976). More recently, however, investments made by private sector interests have begun to dominate. A third special class of producers are those living in villages close to the capital who are under contractual agreement with traders to produce vegetables for the growing European export market. While such arrangements have existed for decades, they have begun to expand recently under the influence of investments made by USAID and the OHVN extension programme in promoting export-oriented vegetable production.

The South-east Portfolio In the *secteurs* of Gouani, Ouelésséhougou and the central and southern parts of Dangassâ (see Figure 3.4), average annual rainfall, can exceed 1000mm, permiting the cultivation of a broad range of crops. Based upon the historical links with the Compagnie Malienne pour le Développement des Textiles (CMDT), which serviced much of this area until the late 1970s, and the continued support from the OHVN extension programme, cotton production remains a dominant feature of the production decisions made by households in this area. Cotton is generally grown of the more fertile and accessible bush fields, and is typically followed by a maize crop to take advantage of any residual fertilizer. To meet household consumption needs, farmers rely upon their numerous local varieties of sorghum, millet, and African rice (*Oryza glaberrima*), in addition to maize. Although millet is of less importance among farm households in this area (due, in part, to intense levels of bird predation), farmers still cultivate short-season varieties, as well as short-season varieties of maize, as protection against years of poor rainfall. A peanut rotation is used in most permanently cropped cereal fields to help combat the emergence of serious weed problems and maintain soil fertility.

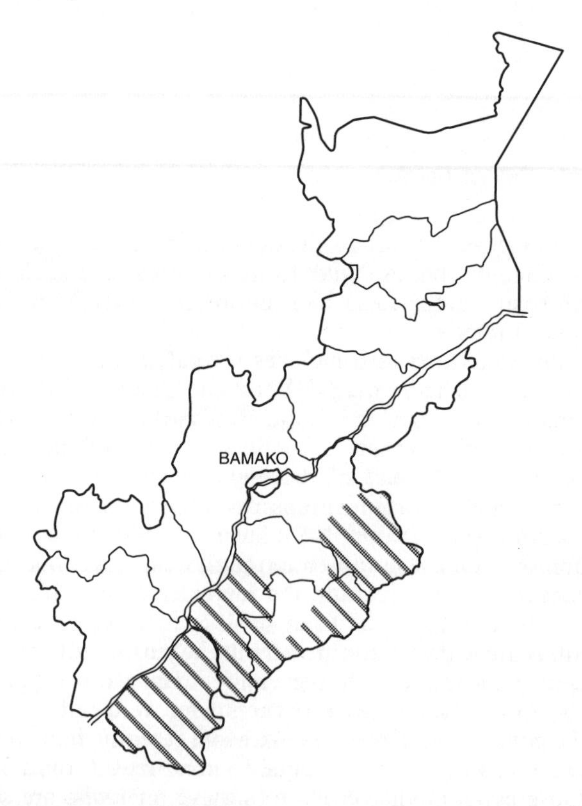

Figure 3.4 *The South-east Portfolio area*

In addition to staple cereals, households cultivate cowpeas, peanuts, Bambara groundnuts, and gourds (calabash, *Lagenaria siceraria*, and edible varieties) in their cropping rotations. Small vegetable gardens, managed by women, produce numerous condiments, while men often maintain a small 'insurance' garden of manioc, as well as an annual tuber crop of yams (*Dioscorea spp.*). Yams have traditionally been an important food and economic crop in communities around the Oueléssébougou area. In addition, many households also plant water-melon as a late season cash crop.

Other economic opportunities include irrigated dry-season tobacco production for those living near the Niger River and other reliable sources of water.[9] Products from the extensive *karité* parks, interspersed with *néré* and the occasional tamarind, *raisonier* (*Lannea microcarpa*), fig (*ficus spp.*) and other species, in addition to 'wild fruits' (e.g. *Grewia spp.; Vitex doniana; Ximenia americana; Zizyphus mauritiana*), are widely used as food and to generate income, especially among women. Bee-keeping is another prominent activity in this portfolio. With a secure and fairly profitable money-making opportunity through cotton production, off-farm non-agricultural activities play an important, but less vital, role in the economic security of households in this portfolio area.

Under growing land pressure, continuous cultivation is becoming increasingly common among many communities. This trend has led to restrictions in women's access to personal fields. In some locations women are able to secure only *usufruct* rights to their husbands' fields during the peanut rotation in the cropping cycle. Women typically utilize some variant of a cropping pattern involving peanuts, intercropped with Bambara groundnut, and over-planted with millet and okra or *dah*. Most villages in this portfolio area are situated near low-lying annually inundated areas, *bas fonds*, where women engage in rice cultivation.

The South-west Portfolio The household cropping systems found throughout most of Bancoumana and Kangaba *secteurs* (see Figure 3.5) are more distinctly divided between lowland and upland fields than in other areas of the OHVN zone. In addition to natural inundated areas, indigenous water-control structures–polders built during the colonial era and drainages improved with check-dams (constructed of rock and wire-mesh gabions)–allow extensive production of aquatic and floating rice varieties (*Oryza spp.*). Upland rice and moisture-tolerant varieties of sorghum are commonly planted in fields leading down to these seasonally flooded areas.

The decline in precipitation since the 1960s has prevented farmers from double-cropping particular fields, and has caused many of the polders and flood areas to receive insufficient water to sustain rice production. As in portions of the South-east Portfolio area, extensive maize planting has taken over areas previously used in rice production. The expansion of cotton production has also partly compensated for the decline in rice cultivation. However, both crops (cotton and maize) are cultivated almost exclusively by men.

Farmers' management strategies include the extensive use of local varieties. Rice producers in particular select from more than a dozen local varieties in responding to both weed pressures and anticipated water depth. In some locations, upland varieties of traditional 'red' rice (*O. glaberrima*), which are particularly tolerant of weeds, are used as the last crop in the rotation when weed pressures are too great for other crops to survive, in much the same way that fonio is used in the north when soils become too impoverished to support any

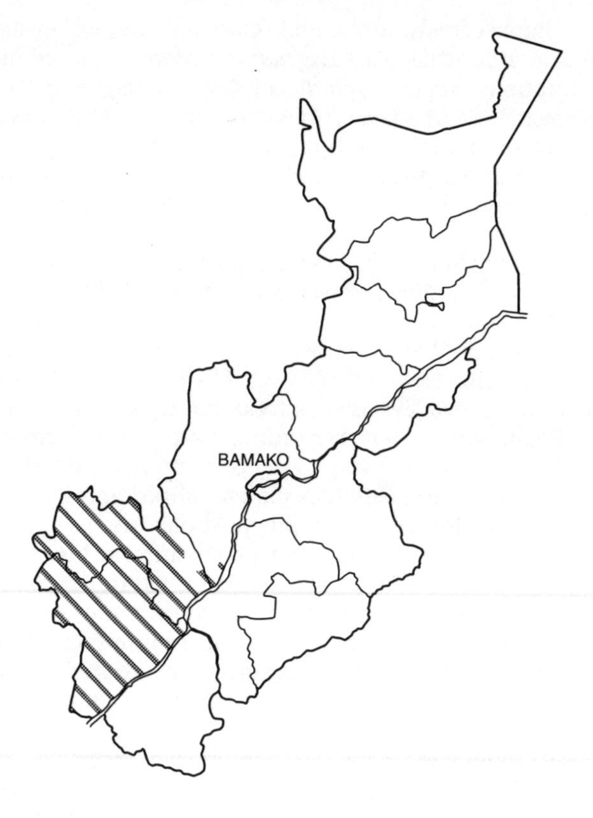

Figure 3.5 *The South-west Portfolio area*

other crop. Other varieties of 'red' rice are used in rattoon culture, selected for their ability to be harvested up to three times during the growing period.

As elsewhere in the southern *secteurs*, Peul herders, driven southward during the droughts, have settled in satellite communities near established villages. In most cases these herders have entered into reciprocal relationships with the farming communities, managing village cattle in exchange for various payments and the right to cultivate land.

Intensely integrated home gardens, based around plantations of mango, citrus, guava, banana, and other perennial fruit species, intercropped with cereal and vegetable crops, are common for many households. These home gardens are typically located near the village, and represent a shift towards a more market-oriented production strategy, as well as a 'retirement' investment for older farmers. Most households manage their own nurseries within these garden plots to provide seedlings both for their own use and to sell. Many of the home garden managers (male) have gained significant grafting expertise from developing their own varieties of fruit trees. The use of *pourghère* fencing around the garden areas, and for entire fields, is widespread. As elsewhere, women use *pourghère* seeds for local medicines and soap production.

In the absence of a strong institutional history of sustained cash-crop promotion, such as with cotton in the South-east Portfolio area,[10] households look to a wide range of income-earning opportunities, from specialty crops such as okra and onions, and to a number of non-farm and non-agricultural activities. Farmers who have developed skill in grafting trees through working in their home gardens are able to secure seasonal employment on the private plantations near Bamako. Commerce along the Guinea frontier and gold digging are among some of the more common and important activities. The allure of high profits in gold mining, in particular, has had major impacts on agricultural activities in some areas. Villagers heavily involved in gold mining often stay in the gold fields until the last possible moment before returning to their villages to begin preparing their fields. Although they may possibly be maximizing their economic returns, their fields suffer as a result.

Summary

The dynamic nature of the physical and social environments found within the OHVN zone requires that farmers maintain a highly adaptive outlook in managing their agricultural activities. Unplanned fluctuations in labour availability, human and animal health, the state of equipment repair and market prices, among numerous other influences, require endless adjustments in farmers' management plans. When set against the backdrop of a highly variable climate, disparate soil fertility levels, as well as disease outbreaks, predation and weed pressures, farmers rarely, if ever, face the same set of annual production decisions. Constraints imposed by growing land pressures require that farmers' decisions increasingly include a longer-term view in the management and use of natural resources.

The need for adaptive management responses and more strategic investment decisions extends well beyond agriculture, and, in fact, permeates the very nature of household decision-making and the allocation of scarce individual and household resources. To moderate the risk they experience, as well as to develop and capitalize on alternative opportunities, farmers spread their resources across a diverse range of income-generating activities, the array of which may change with each season. The widespread influences of certain environmental conditions, marketing opportunities, transportation infrastructure and institutional histories, among other factors, provide households across large areas with fairly similar sets of opportunities and constraints. By examining the pattern of enterprises exploited by households in different locations, as well as the underlying environmental, economic and institutional contexts, at least five, and possibly as many as eight, distinct household livelihood portfolio areas can be described within the OHVN zone.

The concepts of agricultural performance, economic diversification and household livelihood portfolios provide 'tools' useful for both better understanding farmers' behaviour in difficult environments, and in assisting external agencies to better serve farmers and farm households in these areas. By building technical support services that more closely match the realities of farmers' conditions, including their need to maintain an adaptive outlook in their decision-making, external agencies could greatly enhance their ability to provide households with the type(s) of assistance required. The next chapter will explore further the behaviourial dimensions of local agrarian change, examining the processes by which farmers introduce change into their own production systems, i.e., what farmers do and do not know, how they develop new management responses, and how they communicate these to one another.

4 Farmers' Knowledge, Communication and Creativity

Farmers' capacity to enact successful agricultural performances and exploit an evolving range of economic opportunities (described in Chapter 3) is based almost entirely upon the use of their own knowledge, communication channels and creative capacities. In exploring these more intimate aspects of local agrarian change—knowledge, communication and creativity— this chapter begins with a brief description of some of the major attributes of farmers' environmental and agro-ecological knowledge. Findings are then presented on the patterns of local communication and acquisition of materials central to farmer-driven change. A number of social factors are identified that strongly influence both the contexts within which individual learning takes place, and the types of communication channels which different individuals have access to and utilize. The chapter concludes with a look at farmers' creativity, the nature of local experimentation and provides some brief examples of the ways in which farmers are attempting to address major concerns in their production systems. It is argued that a clearer understanding of these internal dynamics of the local change process is critical if the effectiveness of formal research and extension services in drawing upon, interacting with or contributing to endogenous agricultural development processes is to be improved.

Farmers' environmental and ecological knowledge

As one of the central elements in successful agricultural and natural resource management, farmers' environmental and ecological knowledge warrants special attention. Such knowledge not only enables households and individuals to exploit the various resources available to them, but it also serves as the basis upon which households and individuals make decisions to adopt, adapt, reject or develop new management tools. While many of the biological and economic attributes of traditional farming systems have been documented,[1] and are becoming better understood through the growing body of agro-ecological literature focused on farmers' production systems,[2] much less is known about the specific details of farmers' physical and biological knowledge and how this knowledge is used in making key management decisions (e.g., Sharland, 1991) (see Box 4.1).

As the example in Box 4.1 illustrates, farmers in the OHVN rely heavily upon their detailed environmental and ecological knowledge in making important management decisions at the major junctures in the cropping cycle. Observations of weather patterns and constellations, the behaviour of certain plant, animal and bird species (as well as specific individuals), are used in forecasting the change in seasons and making decisions on when to begin land preparation and planting. This same knowledge is also important in managing specific resources on a longer-term basis. Using a range of indicators farmers differentiate between, assess the productive potential of, and monitor changes in various environmental niches, such as observing indicator species and the composition of vegetative communities as a proxy of soil fertility. Detailed knowledge of the requirements and tolerances of individual crops and varieties allows farmers to mix and match the use of specific varieties in different physical locations and at different times.

Box 4.1 Excerpts from an interview with Owenemen Diarra, 27 May 1991

The way the rainy season approaches in these parts, there are signs that we use which let us know that the rains are not far off. What things? Well, there is a certain grass which grows in the bush lands called *musowa*. The hot weather that comes before the rains makes this grass sprout new leaves. The cattle feed on these leaves. This grass sprouts before another that we use to make granaries from, called *cékala*. When you see that all of the *cékala* has sprouted, the rains will have arrived. Whatever you do, you should be planting then. If you've planted by then you won't get left behind. That's what we know about that, at least that's what comes to mind today...

...What we know about birds, well, that's something that everyone knows. So if I skipped over that in the first place, it's because it's common knowledge. There is a bird that comes around here called *baninkono*. When you see the *baninkono* in the treetops, you should take your axe and go clear your field. God willing, you should be able to finish before the rains have arrived. If it turns out that you've cut the trees from your field by the time the *baninkono* arrives, then you'll be able to get a good burn... But it's just a bird, and about all you can count on is that when it comes the rains aren't too far off. In order to be more precise, we watch the way it acts... It's when it comes to build its nest that your planting should be done. You need to see the *baninkono* gathering millet stalks to take them up in the trees to build their nests and settle down. When they've built their nest, you'll find us planting.

...[T]here is another bird with a white chest that comes from the river when the rains are about to arrive. Quite a lot of them show up here. When we see them around, we know that it's time to get to work. The old men tell us that when we see these birds we should tell the kids that it's time to get to work. If you need to clear your field, you should take your axe and go clear your field. If you need to hoe your field, you should take your hoe and go work your field. When these birds start flying around in the tree-tops our young blacksmiths, who make our tools—our hoes with the curved handles, and especially the blades for them—they drop whatever else they're doing and get to work making our farm tools. If this is done as it should be, we'll be able to buy the blades and put them in without getting behind in cultivating... (McConnell, 1993: Appendix A).

While much, if not most, of this type of environmental and ecological knowledge is culture-, and even location-specific, a number of aspects of farmers' indigenous knowledge can be used in highlighting distinct patterns in the nature and structure of local knowledge systems in general. A better understanding of both the major strengths and limitations of farmers' knowledge and perceptions will be an important starting point in helping to guide the efforts of outsiders to engage local resources in formal research and development efforts.

Farmers' management of micro-environments

One of the hallmarks of agriculture in diverse and risk-prone environments is farmers' use of their detailed agro-ecological knowledge in exploiting what Chambers (1990) refers to as 'micro-environments.' As elaborated in Chapter 2, neither all farmland under the control of a community or household, nor all land within a single field, has equal agricultural potential. This fact requires farmers to adapt their practices to fit the prevailing conditions on a field-by-field basis. In doing so, farmers select crops and individual varieties to match soil types and moisture regimes (cf. Matlon, 1980; McConnell, 1993; Moseley, 1993; Stoop, 1987b; amongst others), levels of soil fertility and weed infestation (cf. Richards, 1986), and for farmers involved in aquatic rice cultivation, anticipated or managed water depth (cf. Harlan and Pasquereau, 1969 Richards, 1985;). In the Central and Southern Portfolio areas, farmers use the humid conditions found beneath *karité* and *néré* trees to plant crops of tomatoes and peppers (*piment*). In the Far North these same micro-sites are at times used to support limited plantings of sorghum, while the surrounding areas are planted to millet (cf. Cross and Barker, 1991; SPORE, 1993). Small corner plots with unique hydrology, hedge rows and boundary areas are all used for both specific production activities and as reservoirs of wild plants important to various household activities. Specific sites, such as perched basins on the otherwise uncultivated hardpan plateau, which retain water during the rainy season, are used to support limited rice production, while the nutrient-rich and moisture-retentive soils near termite mounds are used for the cultivation of specific crops and varieties. Within the context of the household production system, farmers' knowledge about, and their use of, these and many other micro-environments provide an important set of opportunities that farmers exploit on an individual and household basis in meeting their economic and nutritional needs.

Farmers' descriptive and taxonomic knowledge

Far from representing a loose collection of 'facts' and observations, much of farmers' environmental knowledge is organized into a number of descriptive classification schemes. While the extent to which the various systems represent structured taxonomies varies, there is little doubting farmers' differentiation between and among the major classes of resources important to their livelihoods (e.g. land forms, soil types, plant species, animals and others). Not surprisingly, farmers' endogenous classification systems have been among the most closely studied aspects of local environmental knowledge in Mali, the region and Africa in general.[3] These studies, when combined with farmers' explanations and additional observations on their behaviour in the case of the OHVN, allow a number of points to be raised regarding the 'cultural embeddedness,' content and structure of endogenous descriptive systems.

At the broadest level, endogenous systems of classification, and the individual objects contained within them, often form part of, or are incorporated into, a larger cosmological or religious belief system (e.g. Haverkort *et al.*, 1996; Millar, 1996). In the case of the Bambara creation myth, for example, the combination of four base spiritual elements (represented by earth, air, fire and water) 'form a framework in which an identification of 968 forms of animal life and as many or more plant forms are made' (Zahan, 1979:43).[4] Such complexes of traditional belief often give rise to a number of specific behaviours that have important implications for agriculture and agrarian change, such as the widely observed prohibitions concerning the avoidance or socially differentiated use and consumption of specific resources or foods, and the more general overarching beliefs in the ebb and flow of seasons, cycles of birth, death and renewal (e.g. Bâ, 1972). Endogenous world views are often used to explain the occurrence of specific events, such as drought or illness, as well as on-going processes, as in farmers' interpretation of highly eroded sites and exposed areas of lateritic hardpan as the active work of spiritual forces. To mediate between the physical and spiritual worlds, local religious figures are often consulted, or discrete rituals performed, before engaging in certain acts, such as preparing one's fields, planting or felling a large tree. The existence of such social-organizational structures and protocols are important reminders of the physical and spiritual unity often perceived in local cosmology (e.g. Bâ, 1972). While often deeply felt at a personal level, the influence of the traditional belief system and its associated practices is not equally spread, and appears to be eroding in importance in many locations, especially among the young.

Despite the attractiveness of a single, well-integrated, overarching, explanatory system, the linkage between the larger religious, or cosmological, vision and the individual systems of classification (e.g. crop varieties, soil types) is not always evident or well defined. In the case of soils, for example, while the earth (soil in general) is a central part of the Bambara creation myth (Zahan, 1974), individual soil types do not appear to be accorded any particular religious significance. Soils tend to be given names for features in the landscape where they are found, or for specific attributes, such as their colour or texture, and are managed as farmers see fit. A number of additional attributes of local descriptive systems can be highlighted through taking a closer look at the example of local soil classifications, as well as farmer behaviour and the extensive research that has been conducted on indigenous soil typologies.

One of the more apparent and important features of indigenous soil typologies is that they are oriented towards practice. 'Farmers are most interested in features of topsoil... [that]... influence important management decisions' (Sikana, 1993:15; Pawluk *et al.*, 1992). For farmers in the OHVN zone, the most prominent soil-borne influences on crop production are soil fertility and moisture retentiveness. These two features are closely associated with soil texture and structure, which, along with a soil's position in the landscape and colouration (indicative of the parent material), are among the most important characteristics used by farmers in differentiating between soil types (e.g. Aubert and Newsky, 1949; Bradley, 1983; Dabin, 1951; Diallo, 1991; Moseley, 1993; Sikana, 1993; Warren, 1992).[5] Macro- and micro-soil pore structure largely determines the moisture retentiveness; soil structure and texture are also related to the proportions of clay, silt and sand, and the presence of soil organic matter. A high clay content in the surface horizon (especially montmoullorite-type clay) represents a higher capacity to hold nutrients in the soil profile where plants can access them. By contrast, in highly leached soils, and those without appreciable clay content in the upper horizons (or clay of the appropriate type), the nutrients are concentrated in the soil organic matter, thus making soils with the highest organic matter content the most fertile. A texture-responsive soil taxonomy, therefore, allows farmers to differentiate among soils based upon an assessment of those properties that reflect the most limiting factors affecting production.

A second important attribute of indigenous soil taxonomies is that they also tend to be based upon the more easily observed, or culturally important, characteristics of the resource. Research conducted in the region on indigenous typologies leaves few, if any, doubts over the ability of indigenous taxonomies to demarcate soils with distinct structural and chemical properties (e.g. Bradley, 1983; ICRISAT, 1989; and others). Some soil attributes, however, such as acidity and aluminum toxicity, which are widely prevalent in the OHVN zone (where most soil pH ranges from 4 to 6, e.g. Diallo and Keita, 1992; Moseley, 1993) and can significantly affect the yields of many crops, are not always well recognized in indigenous classification schemes (e.g., Bradley, 1983). Thus, even though such soil features are agronomically important, they are also relatively 'invisible' in that they are not easily linked to observable characteristics of the soil type, and their influence on crop performance is easily masked by other more obvious stresses (such as moisture availability), as well as individual crop tolerances.[6] This phenomenon, as explained by Bentley (1992) in the context of a study of farmers' entomological knowledge in Central America, is based upon the following 'two principles: ease of observation (or conspicuousness), and cultural importance (or perceived importance)' (Bentley, 1992:10). Acidity certainly ranks low on observability, particularly when it tends to be a general condition of the majority of soils in an area, as is the case in the OHVN, and may even rank low on perceived importance if it cannot be separated from other influences and directly linked to crop performance. Bentley (1992) notes that insects of major importance to agriculture, such as parasitic wasps that help control populations of serious pest species, are accorded little attention by farmers because their actions are inconspicuous and rank low in perceived benefits. The important lessons to be drawn from these observations concern the limitations of local knowledge systems—'what farmers don't know can't help them' (Bentley, 1989; 1992)—and the genuine need for integrating the informal and formal knowledge systems.[7]

A third attribute of local systems of soil classification is that they tend not to be organized into highly regimented structures, where all objects are measured against some fixed criterion, or absolute scale, and ordered in ascending or descending fashion. Rather, categories are ordered along comparative lines, where objects are defined by their relation to one another (e.g. Aubert and Newsky, 1949; ICRISAT, 1989; Sikana, 1993). In fact, the criteria used to differentiate between objects may change from farmer to farmer, and from object to object. Bradley, in his study, analysed collections of soil samples, classified by farmers into indigenous categories, on 13 attributes (chemical, structural) measured through laboratory testing. Although some degree of latitude was present in the placement of soils into the different categories by each farmer, the more significant tendency was for the specific attributes of the different categories themselves to shift, moving towards the dominant soil quality of the area where the samples were collected (Bradley, 1983). Thus, in areas with predominantly sandy soils, all categories would contain a higher sand content than the same soil categories found in regions with an overall lower sand content, supporting the observation that indigenous classification systems are 'relative and site-specific rather than absolute and universal' (Sikana, 1993:16; cf. Aubert and Newsky, 1949; ICRISAT, 1989).

Finally, while a basic understanding of local taxonomic or descriptive systems of classification are, at a general level, broadly shared among farmers, individual farmers' knowledge of the more detailed attributes of objects within these classification schemes are largely based upon personal experience. In other words, the breadth and qualitative depth of individual knowledge is most strongly influenced by the range of personal experiences involved in managing and using different resources in specific ways, and the length of association with the resource. In the research conducted on local soil classification, the most detailed taxonomies have generally been recorded by researchers who contacted the largest number of farmers within a locality. This observation is substantiated in part by research conducted in Northern Zambia (Sikana, 1993), where farmers were on average able to describe in detail at least three soil types, generally those that they farmed. Beyond this, the level of knowledge became increasingly vague, with farmers being able to name additional soil types, but less able to describe their specific attributes. Such an experiential-based view of local knowledge accumulation emphasizes the personal and evolving nature of individual knowledge, which begins early in childhood as children begin to develop a 'cognitive map' of the village's lands while carrying out daily chores, such as gathering wild crops or minding the family's livestock (e.g. Katz, 1991; cf. Millar, 1996). This finding on the importance of differences in individual knowledge has significant methodological implications for recording and understanding local knowledge systems, as well as for establishing working partnerships with farmers.

Farmers' knowledge of agro-ecological processes

As anyone working closely with farmers is quick to realize, farmers' agricultural knowledge extends well beyond the static characteristics of descriptive systems, and includes an appreciable, often highly intimate, understanding of important biological and physical processes. Farmers are well aware, for example, of the fluctuating nature of soil fertility, depleted through cultivation and restored by crop rotations involving nitrogen-fixing legumes, the retention of crop residues, addition of organic matter, chemical fertilizers, fallow and deposition of nutrient-rich 'dust' during the *harmattan*. Farmers often link a decline in soil fertility to specific changes in their management systems, such as the reduction in fallow length, increased use of the plough or long-term use of inorganic fertilizers. In the major cotton-producing areas of the South-eastern Portfolio, farmers have begun to recognize the negative impact that cotton production has on their soils in terms of increased soil erosion (cf. Moseley, 1993). Farmers often describe these processes in terms of the 'tiredness' or 'oldness' of land with declining nutrient stores (or the appearance of 'wrinkles,' caused by erosion as another sign of a site that has grown 'old,' e.g. Jungerius, 1985). While at times such statements have been misconstrued as evidence that farmers are ignorant of the effects of soil erosion, recent studies (e.g. Reij, 1988; 1991; Reij *et al.*, 1996), as well as archaeological evidence from sites across Africa (e.g. AZANIA, 1989; Sutton, 1990), indicate that farmers have a rich history of using and adapting soil- and water-conservation measures that dates back several centuries.

As in the case of soil typologies, much of rural people's knowledge regarding biological and physical processes appears to be based upon those features that are more readily apparent, e.g. in the form of a causal relationship, and linked to the principal limiting factors in the production systems. Women farmers in the Southern Portfolio areas, for example, when asked if they recognized any positive benefits in their intercropping pattern (a common, dense association of peanuts, Bambara groundnuts, millet and okra or *dah*), lamented the lack of benefits, stating that they only used intercropping practices because of the scarcity of land—they would need more land in order to grow the same amount of each crop separately. While they did not view intercropping in itself as beneficial, it is obvious from their response that these women made explicit use of the 'over-yielding' potential of intercropping systems.[8] In this instance, however, the benefits of intercropping were not interpreted in terms of yield, but rather in terms of the women's most limiting resource—land. In contrast, male farmers in the same area, who are less constrained by land availability (because men control land allocation), referred to the benefits resulting from certain associations and rotations involving cereal crops and legumes in terms of yield, while making no mention of the 'land saving' ability of intercropping. Similarly, farmers in the more arid Far North Portfolio area reported that their common practice of planting beans in millet fields did not noticeably help or hurt the primary millet crop, but if there was sufficient rain, the beans would provide an additional crop as well. In this case moisture from a limited and highly uncertain rainfall, and not soil fertility or land availability, was the principal factor around which farmers based their judgements and interpreted observations. The bean populations in this type of intercropping arrangement were generally far too low for the legume's nitrogen-fixing abilities to benefit the cereal crop significantly, or for the beans to compete strongly for soil moisture, yet with sufficient rainfall both crops could prosper and provide a yield. These brief examples illustrate farmers' basic understanding of what has become known in agro-ecological literature as the competitive and facilitative production principles of intercropping (Vandermeer, 1989).[9] Even though farmers may not perceive, or explicitly apply, these aspects of their knowledge as management principles, apart from the context of their specific production systems, this type of convergence between farmers' knowledge and that of researchers and extensionists does provide an entry point for farmer–researcher collaboration in addressing problems of immediate concern.

Communication and the social differentiation of knowledge

Fieldwork conducted in this study revealed what can perhaps best be described as a 'topography' of local knowledge, where individual producers possess heterogeneous sets of knowledge consisting of elements of both commonly held 'traditional' wisdom and individual understanding that differ in both quantitative and qualitative terms, within as well as between villages. Individual differences in knowledge are often based upon the social importance of gender, age and ethnicity, and are affected by other less apparent, yet equally important, influences involving kinship ties, religious affiliation and wealth (cf. Swift, 1979). At any one time individuals participate in a number of different, overlapping social networks from which they receive, and to which they contribute, information. These different associations not only influence the type of information to which individuals have access, but also serve to delimit the type of experiences and opportunities through which individuals can observe and interact with others. While elements of each individual's ascribed and achieved social status help to define the most active channels of interpersonal communication used in the diffusion of new information, friendships and personal affinities can be seen as an important 'social lubricant' in helping to facilitate the exchange of information and material along and, less often, across socialized lines of communication. Specific locations, such as public spaces and markets, and certain activities, such as group labour and other events, provide the physical and social contexts within which the exchange of information between individuals often takes place. Certain types of exchanges, such as those that occur at local and regional markets, serve to bridge both geographic and social distances, and are critical in speeding the movement of information. Together, these and other forces help to define the range of personal experiences and communication linkages that are most important in understanding the patterns of local knowledge differentiation and pathways of information exchange.

Families

The family and larger kinship groups found in Bambara society constitute an individual's most important set of social relations.[10] The family structure not only represents a fairly independent, self-replicating social unit that affects the distribution of knowledge within the larger community, but it also serves as the primary context within which other processes of social differentiation (e.g. gender and age) are acted out. Through direct instruction and facit modelling, young girls and boys receive the greater part of their respective role training from the adults in their immediate families (e.g. Cross and Barker, 1991). The decisions made by the household head, and norms of social conduct reinforced by others within and outside of the family (e.g. Grosz-Ngaté, 1989), help to shape many of the subsequent experiences and, consequently, much of the specific knowledge acquired by individuals within the family. While many of the variances in familial behaviour ultimately reflect differences in the financial stature, interpersonal relations and production objectives of particular households, others reflect the customized practices and efficiencies developed within different family groups (cf. Pottier, 1994; Millar, 1996).

Despite several studies in the OHVN zone that have commented on the break-up of the family unit (e.g. Becker, 1990; Gnägi, 1991), farmers generally report that their families are one of the most important sources of information and new genetic material available to them (see Table 4.1). This includes both the immediate, resident family unit and their marketing forays and seasonal labour sojourns, as well as the occasional visits to and from relocated family members where exchanges of information take place (cf. McCorkle *et al.*, 1988). Both women and men farmers in the zone reported a number of successful exchanges, mainly of genetic material, that occurred through long-distance familial ties. In Bambara

Table 4.1 ***Relative importance of various sources of information****

General sources of information and material	Responses Men (n=11)	Responses Women (n=10)
Research	4	1
Extension	3	2
NGOs	3	–
Radio	4	1
Elders	7	5
Family	6	5
Marabout	2	–
Group members	1	7
Friends/Neighbours	6	1
People at Market/*Commerçants*	5	1
Others	3	–

* Responses are drawn from group interviews (11 male extension groups; 10 women's groups).

society, marriage traditionally occurs between individuals from different villages, with the wife taking up residence in her husband's natal village.[11] Visits by married women to their village of origin was cited as another very important pathway of information transfer. Studies in the Sudan have found similar patterns of information exchange (e.g. Coughenour and Nazhat, 1985; Nazhat and Coughenour, 1987), noting that close and distant relatives are one of the most important initial sources of information about new varieties, particularly for women, and constitute the most important source of information for both sexes in making planting decisions involving new varieties.

At times farmers may delay telling others outside of their immediate families about new varieties or techniques, thus slowing the inter-household transfer of varieties within a community. In the case of new genetic material this is reportedly done not only to save face, if the variety fails, but also to keep others from begging seeds that are in short supply. This delay tactic may also enable households to gain a year or two advantage if the new variety proves superior (cf. Nazhat and Coughenour, 1987). In terms of more general production techniques, farmers may actively attempt to safeguard family 'secrets' (e.g. Pottier, 1994; Millar, 1996). When exchanges do occur, the local flow of information is driven primarily by experience; that is, individuals, and consequently their kinship groups, who have the greatest experience with an innovation tend to serve as the principal source in diffusing the new technology to others in their area. Because each kinship group maintains different channels of communication, especially with regards to extra-village contacts, their relative importance in acquiring and diffusing new technologies is continually changing.

Gender

As would be expected, group discussions revealed significant differences between women's and men's knowledge. These gender-based differences in knowledge are tied to two major themes: the specialization, or genderization, of specific household production activities, and the presence of resource constraints that are gender specific. As observed in the brief discussion on the organization of household productive activities in Chapter 3, most agricultural activities are divided along gender lines under the ultimate direction of the eldest active male, with the eldest active female serving to co-ordinate and organize the activities relegated to the household women. The exact structure of the gender-based division of labour, however, is by no means consistent among all communities, or across all households within a single community. (In fact, in instances where there are no strongly socialized norms of behaviour, the division of labour within the household depends as much upon interpersonal relations and negotiations, as on any broader cultural trends).

In most areas of the OHVN zone, for example, rice production is an activity carried out predominantly by women. Yet in at least one village in the South-west Portfolio Area, rice cultivation has evolved into an exclusively male-dominated activity. In this village, farmers stated that the scarcity of good rice land, aggravated by the decrease in rainfall during recent decades, had led to women being denied access to these areas by the hierarchy of village males. By contrast, in neighbouring villages as little as 15km away, which had suffered equally from the shrinking land base, the cultivation of rice had remained solely a women's activity. Men in these other villages replied that due to the decline in rice producing areas, the growing of rice was not profitable and, therefore, was of no interest to them. In areas further to the south, where rice farming has played, and continues to play, a more significant role in the household production systems, and where rice land is more abundant, both men and women continue to be actively engaged in rice cultivation. Not surprisingly, group discussions in each of these different locations revealed how the depth of knowledge held by women and men on such things as the characteristics of different rice varieties and the importance of different rice pest and weed species varied according to their degree of involvement in rice production. This process of gender-based task and knowledge differentiation is visible in most aspects of the household production system, where the cultivation of specific food and cash crops, the collection of wild fruits and leaves, and activities such as the processing and preparation of food are assigned to different gender and age groups.

Labour, and increasingly land, are among the most important resource constraints influencing gender relations in the OHVN zone. In most locations, younger women are prohibited by the male household heads from diverting labour from household duties in order to cultivate their own personal fields. However, once women gain access to land, and the use of their own labour, they are able to exercise complete autonomy in making planting decisions, varietal selection, field management and granting minor land concessions to young girls (who perform various favours for them, such as bringing food out to the field or helping with the weeding). While women in one area reported receiving all of their seed stock from their husbands (i.e. family ties), this was not true for the majority, who reported locating, conserving and exchanging varieties separate from their husbands. In an increasing number of locations in the South-eastern Portfolio area, population growth and a decline in arable land due to the drop in annual rainfall levels has resulted in women being able to secure annual *usufruct* rights to their husband's fields only during the peanut rotation in the cropping cycle. Thus confined in their choice of crops and their degree of control over the land, women's opportunities to experiment with and implement alternative management practices have been severely limited, while their knowledge and expertise in implementing a single-season peanut production strategy has increased. Ownership patterns and access to other resources and opportunities (e.g. animal traction and formal credit) further influence gender differences in knowledge accumulation.

Gender also serves as the basis for important lines of communication. As noted in the discussion of kinship ties, the socialization patterns of women and men in Bambara society can act as an important force in knowledge differentiation. After the age of 8 to 10, the instruction of children within the family occurs almost exclusively along gender lines, as children begin to accompany their parents into the fields—daughters with mothers, and sons with fathers (e.g. Cross and Baker, 1991). The behaviourial expectations of the ascribed gender roles are further reinforced by those outside of the family. As they gain skills, children become increasingly involved in a number of gender-specific activities and work groups. Their increased responsibility for daily chores brings with it a growing participation in a broader range of communication channels. Women, for example, exchange information and experiences in the course of performing many regular domestic and agricultural tasks, such as getting water, gathering firewood, and while walking to and from fields, local markets and other destinations. Young girls, assisting older women in their personal plots, receive numerous 'lessons' from their mentors/benefactors in performing different field operations, and in this way learn a great deal about specific agricultural techniques that they will be able to apply on a larger scale in their own plots once they gain access.

For women, their village association meetings and group labour activities are among their most important periods of information exchange. This is quite different from men, who reported almost no exchanges of information or materials with other members of their extension groups, but who cited a high number of such exchanges with 'friends and neighbours' outside these formal associations. Unlike the all-male extension groups, formed through the encouragement of the extension service, most women's groups represent indigenous social organizations founded upon mutual need and maintained by close personal ties. Many, if not most, of a women's close friends are members of one or more of her different group associations, where the group activities provide the context within which much of their interaction takes place.

Other types of gender-based associations at the village level also provide important opportunities for exchanging information. During and after the harvest, women often form *ad hoc*, multi-household work groups. The presence of males beyond the age of infancy during winnowing, for example, is strictly forbidden (Lewis, 1979), providing women a period in which they can freely discuss events and activities that would otherwise be constrained by the male-dominated social hierarchy. The male age-set *tons* represent another form of gender-based organization. In hiring themselves out for field labour, *tons* members are exposed to a number of different production styles and gain valuable information from their peers and the farmers in whose fields they are working. Farmers who contract group labour often submit at least partial 'payment' for these services in the form of advice or wisdom relating to agriculture (McConnell, 1993). In a study on traditional apiculture practices in the Oueléssébougou area, Gnägi (1992) found that when honey gatherers (all male) congregate and return to the village at the end of the day, they engage in personal discussions that range from experiences with their hives to commentaries on the state of the fields through which they are passing, as well as personal problems (pers. com. Gnägi). These types of moving commentaries are common (e.g. Hoskins, 1994; McConnell, 1993), and in one sense represent an on-going indigenous analogue to the Rural Appraisal techniques involving farm visits or transects.

Age

The influence of age on differences in personal knowledge and communication pathways is another culture-wide phenomenon. Wisdom gained from long years of experience in agricultural performances over a range of conditions is, in general, widely revered in Bambara culture (cf. Rosenmayr, 1988). Technical expertise, 'the know-how which, in rural Mali, is traditionally acquired gradually over time and which the old are suppose to have the most of' (Jones, 1976:285; cf. Meillassoux, 1960), has been suggested as the elderly's essential contribution to production, and hence the basis for their position of authority. Not surprisingly, women and men farmers throughout the zone view village elders as among their most important sources of information (see Table 4.1). At times, older, more experienced male farmers, 'retired' from active cultivation in the family fields, give advice to young farmers who are not part of their immediate kin group, forming apprentice-type relationships (cf. Millar, 1996). As noted earlier, elder women perform the same role in their relations with younger assistants in their personal fields.

In Bambara society, the first born, or eldest children, are reportedly given more detailed agricultural instructions by their elders (McConnell, 1993) and, in turn, are responsible for much of the education and development of their younger sisters and brothers. Older children are also the first to assume positions of leadership and responsibility within the family as they mature. In this way, age results not only in quantitative knowledge differences, based upon levels of experience, but also includes elements of qualitative difference, due to the nature and level of detail contained in the information passed on by members of one generation to the next.

In general, women expressed greater respect for their elders as important sources of agricultural information and advice than did men. In several villages in the South-eastern Portfolio area, young extension group members, who were exceptionally eager to adopt the 'modern' cotton production practices promoted by the extension service, spoke derisively of their elders' abilities to provide 'correct' agricultural information. At the same time, these younger farmers possessed significantly less knowledge about traditional farming practices and local varieties. This phenomenon, however, was not universal. In neighbouring villages, where farmers were equally involved in cotton production and reliant on external inputs, yet where the relationship with the extension service was strained, no such glowing impressions of modern practices existed; even the youngest farmers were both respectful of their elders and highly conversant in traditional farming methods. This example points to the significant influence that agricultural 'modernization' programmes can have on the local social fabric, and the care that must be exercised when designing information campaigns in order to avoid fostering unintended consequences that may hamper overall development objectives.

Ethnicity, lineage and caste

Ethnicity, often associated with occupational specializations, e.g., agriculture, herding and fishing, contributes to some of the most pronounced forms of social division within the OHVN, and can be seen to significantly affect the differentiation in individual knowledge and local information exchange. The most prominent example of ethnic division in the OHVN is that between groups of herders and agriculturalists. Forced to migrate southwards during the recent cycle of droughts, many groups of Peul herders have since settled in the Southern Portfolio areas and increasingly begun to adopt sedentary systems of crop production. The agricultural efforts of these groups are regarded with some mirth by the resident farming communities as 'novice' at best, replete with numerous 'mistakes.' Yet, the Peuls are entrusted exclusively with the care and management of the agriculturalists' livestock (which constitutes their major form of investment), because of the unquestioned superiority of their animal husbandry knowledge and skills.[12]

Even within agricultural communities, ethnic identity often serves as an important focal point in social differentiation. The social structure of each village, including the origins of the village and hierarchy of each family group, is well known by its members. Despite the cultural similarities between the two dominant ethnic groups in the OHVN zone, the Bambara and the Malinké, newcomers entering a village dominated by the other ethnic group are often discriminated against and excluded from the resident decision-making bodies (e.g. Gnägi, 1991). Such divisions persist over time, becoming part of the socially reproduced *status quo*, and can significantly affect the intra-group exchanges of information and genetic material (e.g. Nazhat and Coughenour, 1987; Weller-Molongua and Knapp, 1995). In one of the study sites, events which took place during the village's formation resulted in a particularly sharp division between the two ethnic groups involved, which made convening any broad-based communal meetings impossible. Gnägi (1991) reports similar conditions in neighbouring areas of the OHVN zone. The existence of such deep-seated attitudes poses serious difficulties to development organizations and cannot be easily overcome.

An overlapping set of influences affecting knowledge distribution and communication is that of familial status, based on lineage and caste divisions. Traditional Bambara society is structured around several castes: nobles, slaves and craftpersons (Diop, 1971; Lewis, 1979; N'Diayé, 1970a, 1970b). Upon entering a village, visitors commonly engage in a form of playful social jousting, tracing their heritages through the established hierarchies. In some communities, membership in one of the remaining active professional castes, e.g. blacksmiths, can serve as the basis for an individual's exclusion from certain activities, such as participation in extension group meetings (although farmers are loath to acknowledge such discrimination). While villages in Bambara society are traditionally established by families of the noble caste (Lewis, 1979), in some instances members of the former slave caste have successfully come to occupy positions of authority, effectively replacing the dominant founding families if no one of the senior age group from the founding family is capable of assuming leadership (Gnägi, 1991). While the chieftainship can transfer to a member of another family and caste group, this transfer may not be permanent and is generally not inherited. As with ethnicity, caste membership is not an entirely negative influence, and can include access to specialized spheres of knowledge, and sources of income, that are not available to other members of the community.

Religion and initiation groups

In many areas of the OHVN, the continued expansion of Islam into former strongholds of traditional religious belief has significantly influenced a number of aspects of individual knowledge, production objectives and lines of communication. For some, local Muslim leaders (*Marabouts*) have become a valued source of agricultural advice, and are especially consulted prior to planting (cf. Swift, 1979:42; Coughenour and Nazhat, 1985; Cross and Barker, 1991). In the historically under-educated rural areas of Mali, Koranic schools constitute another important influence. Individuals attending these schools receive a different type, and often much higher quality, education than is available through the public schools and adult literacy programmes operating in the zone. In addition, Arabic literacy opens up a much broader pool of printed material than does *Bamanan*, taught in adult functional literacy classes. For the devout Muslim, the mosque becomes a focal point of social activity and a significant hub of information exchange. Those attempting to fulfil their obligations of visiting *mecca* must acquire the necessary capital, and in the process greatly alter their household production systems. For the *el Hadji* (the honourary title given to those who have made the journey), visiting new lands has proved to be a pivotal experience. Discussions with these individuals generally showed them to be among the more inquisitive and experimentally minded farmers, as well as the wealthiest.

The fading traditional system of ancestor worship is centred around a series of initiation groups, one of which, the *Tyiwara*, focuses on agricultural performances.[13] Zahan (1960; 1974) notes that while women are excluded from all the other initiation groups, they are allowed to participate in most aspects of the *Tyiwara*, except for discussions 'concerning the significance of the tools used in farming, [and] of the properties of the soil, etc.' (Zahan, 1974: 20). Symbolized by the antelope (who taught the Bambara agriculture), the *Tyiwara* encourages the development of a strong work ethic and agricultural prowess. Yet,

> the sum of knowledge dispensed by the *tyiwara*... [is far more than] ...a sort of vague apprenticeship in the traditional agricultural techniques, and the teaching of some truths about the sun and the earth. This society deals with all that concerns man's food: the creation of man's settlements and the beginning of agriculture, the movement of stars and seasons, the flora and fauna, iron and the technique of making farming tools, clothing and adornment, the knowledge of poisons and, finally, the 'science' of the mastery of celestial fire (Zahan, 1974: 20).

Beyond their function of exchanging information and instilling values, these groups also once served as the basis for inter-village competitions in performing fieldwork, although McConnell (1993) and Moseley (1993) report that this, too, is on the decline.[14]

Wealth

As an achieved status, wealth cannot be easily separated from many of the other social factors influencing the distribution of knowledge and lines of communication. As noted in the example of the *el Hadji*, wealth is an extremely influential factor on the structure and functioning of the household production system. Varying levels of capital/assets allow households to support different production strategies, and consequently provide individuals with different types of managerial experience. Households possessing the means to invest heavily in cash-crop production, more mechanized forms of cultivation, the construction of fixed assets such as wells, in addition to capital-intensive, non-agricultural enterprises, acquire in the process a very different set of skills and experiences.

In the context of village development, wealthy individuals are constantly sought out by external agencies as the perceived opinion leaders of their communities, as well as the most likely candidates for adopting new, often capital-intensive, production techniques. This bias (cf. Chambers, 1983) is clearly visible in the typologies used by independent researchers, as well as the formal research and extension services, who rely upon relative wealth measures (levels of equipment and livestock ownership) to differentiate between households in both their discussions and planning activities (e.g. DRSPR/OHV, 1992c; Koenig, 1986b). Contrary to much of the development rhetoric of working with the rural poor, greater wealth more often than not attracts greater attention from outside agencies, which, in terms of access to information, often means that these individuals are among the first to learn of new technological alternatives.

It is important, however, to note that for their part, those farmers who are currently the more prosperous, are careful not to incur any negative feelings or jealousies within their communities. Not only do wealthier households often depend upon others for labour, but the vagaries of local production, and occasional strings of bad luck, almost ensure that at some point the tables will turn and those households that are now prosperous will cycle through periods of poverty and become increasingly dependent upon the extra-household social safety-net for their survival (cf. Adams, 1993).

Friendships

An important influence on the exchange of information, cutting across all other social dimensions discussed thus far, is that of friendships. Close personal relationships and affinities not only help to accelerate the diffusion of information among individuals of similar social standing, but can also help to bridge the otherwise separate social spheres created by ethnicity, gender and other social constructs.[15] In general, both women and men reported the greatest number of exchanges in information and genetic material between themselves and friends (for women, other members of their women's associations are generally their age-mates and closest friends) (see Tables 4.1 and 4.2). Friendships can even mediate some of the most deeply engrained social divisions, such as those along ethnic lines. For example, in virtually all communities, Peul herders remain socially and physically separate from the agriculturalists with whom they co-exist; their compounds are typically located at the outskirts of the agricultural villages. Friendships, however, established among younger farmers and pastoralists through *ton* membership (e.g. Leynaud, 1966), established inter-family exchange relations and other contacts, suggest that such divisions can be bridged, thereby offering a channel for the exchange of information and genetic material between otherwise socially separate groups.

The exchange of managerial information between friends is of particular significance because, as discussed earlier in this and the preceding chapter, the knowledge involved in agricultural performances is often highly personal in nature, deeply intertwined with household production strategies adopted to accommodate such sensitive areas as labour availability and financial liquidity. A certain degree of intimacy may, therefore, be required before information on specific management practices is easily exchanged. While farmers freely experiment with the information and materials obtained from market sources, they also reported that the greatest number of rejected technologies originated from these less personal contacts. In contrast, new varieties obtained from friends often come

Table 4.2. *Principal sources in the recent exchange of genetic material**

	OHVN extension	DRSPR research	FAO seed multiplication	Markets and traders	Neighbours and friends	Family	Unknown
Millet	2	1	–	4	3	2	3
Sorghum**	9	1	3	3	4	2	6
Maize	3	2	–	1	–	–	–
Rice	–	2	–	3	4	2	–
Peanuts	–	–	–	8	1	–	–
Fonio	–	–	–	1	–	–	–
Most important overall	1	1	–	1	8	4	–

*Responses are drawn from women and men group interviews (n=21) on adoptions made in the 'recent past' (last 2–3 years).

**Most of the adoptions of sorghum from the formal system involved CSM 388, an improved local variety that has been widely promoted in the OHVN since the mid-1980s (SRCSS, 1987)

with specific planting instructions, or as part of a management 'package' involving the variety and specific cultural practices. In the rice-growing Southern Portfolio areas, such packages commonly include information on the proper water depth in which new rice varieties should be planted, as well as soil fertility and the levels of weed infestation that the variety can tolerate. At a more general level, studies from around the world illustrate that friends not only serve as one of the major sources of new information but, along with family members, are one of the most trusted and influential sources in farmers' decisions to adopt new technologies (e.g. Coughenour and Nazhat, 1985; Lionberger, 1959; Lionberger *et al.* 1975; McArthur, 1978; Nazhat and Coughenour, 1987; Ryan and Gross, 1943).

Locations and activities

Locations and the performance of specific activities deserve special mention as they often provide the physical and social context in which many of the exchanges of information and material take place (see Table 4.3). Women identified group meetings, family and communal labour, and a range of sites where they engaged in specific activities (e.g. village wells, local market, mill, *en route* to and from the fields and collecting firewood), as the most important locations/activities in the exchange of valuable agricultural information. Men, on the other hand, reported that the majority of their important contacts occurred while socializing at the village mosque, in public-gathering places—such as beneath the large shade trees or ubiquitous *mirador* (the raised log platforms found in almost every village in the zone)—or in the courtyard or vestibule of the village chief, as well as during periods of field labour. Both women and men noted the particular importance of local and regional markets (cf. McCorkle *et al.*, 1988; Nazhat and Coughenour, 1987).

Village markets provide individuals with the opportunity to meet and share ideas and experiences with friends, acquaintances and more distant members of their kin groups, as well as a wide range of other contacts outside the local social networks. Farmers reported that local and regional markets, and passing *comerçants* (traders), served as one of their most important sources in acquiring new genetic material; peanut varieties from Senegal, and rice and bean varieties from the Gambia and Guinea, were obtained through commercial channels, and were

Table 4.3 ***Important locations in the local exchange of information****

Locations	Men (n=11)	Women (n=10)
Magasin/Storage shed	2	–
Chef's house	3	–
Mirador/Public place	7	–
President's house/Group meeting	2	4
Market	4	3
Family and communal fields	4	4
Other**	1	4

* Responses are drawn from group interviews (11 male extension groups; 10 women's groups)

** Including: village wells, routes to market and the location where specific chores are performed some of the most widely diffused varieties in the OHVN from any source. Merchants in local markets have been responsible for much of the extensive diffusion of farming implements, fertilizers, and to a lesser extent other inputs (i.e. insecticides and herbicides). Some of the larger national and international commercial interests have attempted to expand their influence beyond the market by conducting village demonstrations, sometimes in collaboration with the research and extension services.

In addition to marketing forays, other types of travel—whether for the purpose of seasonal labour, visiting distant kin, attending a funeral or a circumcision, or embarking on a religious pilgrimage—are another important avenue in the acquisition of new information. As one farmer stated, 'anyone who travels' becomes an immediate source of information. Similar to the exposure to new ideas and materials through the market-place, travel opens up a range of opportunities for individual observation, discussion and purchase that can be truly national and international in scope (e.g. Coughenour and Nazhat, 1985; Hoskins, 1994; Knight, 1974; McCorkle *et al.*, 1988). Those with significant tree grafting experience, for example, who have found dry season employment on plantations in the Bamako area, return to their villages with seedlings, in some instances of genetic stock imported from abroad, which they either plant directly or use in their own grafting experiments. Others who have visited Mecca have returned to experiment with fish farming, commercial horticulture and other activities similar to those observed during their travels. In fact, travel in general seems to increase an individual's level of curiosity, as one farmer who had participated in an exchange visit organized by the OHVN extension service noted, 'now whenever I'm on the road or see something different or interesting in a field, I get off my bike and talk to the farmer, or go into the field and look for myself'.

Farmers' creativity and experimentation

Agrarian change obviously involves something beyond the endless transfer and reproduction of a finite set of information and management practices. In fact, there is probably little risk of overstating the importance of farmers' creativity and experimentation to the well-being and ultimate survival of households in highly variable and rapidly evolving physical and socio-economic environments. In terms of processes, indigenous innovations include those arising from both the lessons learned by farmers in making the continual stream of on-the-spot adjustments and adaptive alterations in their existing management systems, as well as the more proactive attempts to introduce desired changes. Thematically, women and men farmers across the OHVN zone were found to be engaged in a wide range of innovative activities aimed at improving nearly every aspect of their farming systems. Examples of farmers' experimentation range from crop and varietal trials, adjustments of planting densities, spatial arrangements and related cultural practices, to experiments on soil fertility management, erosion control, water conservation, agroforestry and the development of new economic activities.

Conceptually, farmers' most common types of innovative activities fall into three general categories: adaptive experiments involving minor adjustments or modifications to established practices, based largely upon the use of existing agro-ecological knowledge; integrative experiments, oriented towards verifying the utility of and integrating new techniques, ideas and materials into the existing production systems; and exploratory experiments, involving the investigation of novel or original ideas stemming from personal insight, desire or curiosity—as one farmer put it, to simply 'see how it works'. Obviously, such categories are not entirely mutually exclusive; there may be a certain degree of temporal linkage between the different types (e.g. exploratory leading to integrative leading to adaptive experiments).

Overall, the improvisational quality of agricultural performances in the region, described in Chapter 3, casts a definite influence on all types of farmers' experimental activities. At one level, farmers in the OHVN zone can be seen to employ many of the same logical steps used by formal researchers in conducting their integrative and exploratory activities (i.e. screening new genetic material for desired characteristics, identifying suitable growing environments and integrating new varieties into the existing cropping systems) (e.g. Potts *et al.*, 1992). However, due to the highly variable nature of the physical environment, farmers generally do not treat their experiments as 'fixed' investigations of discrete research questions; rather, they allow the conditions of the particular season and their own managerial knowledge to guide the investigative process, changing the nature of the query and 'treatments' as the season(s) unfolds (e.g. Stolzenbach, 1994).

Farmers' willingness to discuss their experimental activities varied widely, from obvious pride in their accomplishments to great reluctance, acknowledging their experimental activities only after examples were stumbled upon in the course of field visits. In villages that had hosted formal on-farm or pre-extension trials of the DRSPR research programme, some of this hesitancy may have been due to farmers' feelings of inadequacy, of not doing it 'right' in comparison with their perceptions of formal research procedures. However, other aspects of farmers' apparent hesitancy may be explained by the deeper influences of semantic and conceptual differences, compounded by methodological issues. In Bamanan, for example, there are several ways of referring to experimental actions, including: *i bila a da la*, to take an action to see if it is successful; *lajè*, to try something (an object) to see if it is any good (Bailleul, 1981); *shifleli*, 'doing something and looking at the results' (Stolzenbach, 1997:40); and *kôrôbôli*, the action of distilling the significance or sense of something (Belloncle, 1979). These expressions, as used by farmers (and understood by outsiders), may not capture the same set of attributes that are implied by Western uses of 'trial' or 'experimentation.' Thus, the example reported by Stolzenbach (1994), where a farmer responded that an apparently new planting practice used with a variety he had adopted in the preceding year was not considered an experiment (*shifleli*) because, in his words, he already 'knew the variety' from the preceding season, is as telling about how farmers may perceive that type of activity as it is about the researcher's interpretation based upon the use of a single descriptor (*shifleli*). Using a range of methods (descriptors), farmers in the OHVN readily reported experimenting with a large number of new crop varieties during the group village interviews. However, the subsequent adjustments in planting densities, use with different soil types, intercropping patterns and related management practices were rarely offered as examples of experimentation in these group sessions, despite the fact that, based on past performances, farmers generally knew exactly what adaptive changes they wanted to make at least one season in advance. Examples of these more intimate, second-generation, management-based, experimental activities did not begin to emerge until individual field visits were made, and discussions focused on field histories and the evolution of management practices. One clear message that arises from these observations concerns the special attention that needs to be paid to methodological issues in conducting further research into farmers' experimentation.

As described in previous sections, farming under diverse and risk-prone conditions is an adaptive process requiring farmers to respond to high levels of variability in their physical, as well as social, environments. Under such conditions farmers experiment through the necessity of making changes, the desire to change in order to overcome perennial difficulties and improve conditions, as well as to explore the possibilities that more speculative changes may offer. Influences such as periodic 'shocks' in market prices, the level of institutional support, or sharp, yet enduring fluctuations in rainfall patterns (all of which occurred during the period under consideration in this study), create immediate, often well defined, needs and opportunities that serve to accelerate the rate of change in local production systems for entire groups, not just individuals. Other influences, such as farmers' inherent inquisitiveness and the basic forces of natural selection (i.e. genetic evolution in crop, weed and insect species) (Biggs and Clay, 1981), as well as longer-term trends in land availability and soil fertility levels, ensure that a certain background level of more sporadic change will always be taking place.

The remainder of this chapter briefly illustrates a few examples of the type and nature of farmers' experimental activities as they attempt to improve both the efficiencies of their production systems and their ability to manage and control the physical environment. Other examples, as well as the cumulative impact of farmers' innovative achievements, will be the focus of Chapter 6, which takes a broader review on the relative contributions that the endogenous and exogenous systems have made to local agrarian change. For the present discussion, however, it is important to bear in mind that the selected examples presented here represent only a thin slice, in terms of time, location and theme, of the total wealth of innovative activities occurring within the informal sector across the OHVN zone. The full range of experiments and adaptive adjustments being carried out by farmers must, in reality, be multiplied across the approximately 800 villages and more than 40,000 households within the zone. Because these changes are diffuse and generally small, compared with the more narrowly defined and episodic contributions from external agents, it is easy for external observers to underestimate the significance and cumulative effect of the informal evolutionary processes. Nevertheless, when viewed over time, each instance of farmer-based innovation and adaptation represents a vital contribution to the incremental advancement of the existing farming systems. While some farmers are more adventurous than others, and some more systematic, the fact remains that virtually all farmers experiment. If they did not, little agriculture, and few people, would currently be found within the region.

Cropping systems

Farmer interviews and field visits revealed that the highest level of experimental activity occurs in the areas of varietal screening and the adaptive adjustments that farmers make in their cropping systems. As a first step in familiarizing themselves with a new variety, farmers typically conduct screening trials in small test plots established in favourable locations, such as within home gardens or more fertile portions of larger fields. These small screening plots serve both as a first test of the variety, and as a way of multiplying seeds for planting on a larger scale (farmers typically buy new varieties in small quantities, or exchange them with others on a measure for measure basis).[16] Although not the preferred practice, new varieties may be planted directly at the field level in cases where farmers are forced to replace much, or all, of their seed stock, due to drought or other losses, or when a sufficient quantity of a new variety has come from a highly trusted source. Having advanced to the field level, farmers may continue to make adjustments regarding where the variety is planted (if they feel that it would be better suited to a different soil type or moisture regime), as well as to initiate the process of either integrating the new variety into the existing cropping pattern, or changing the standard practices slightly to better exploit the desirable characteristics of the new variety. If a variety does not perform to expectations, shows particular promise, or introduces new management problems, then it is generally rejected, as was the case in farmers' trial and rejection of short-season sorghum varieties introduced into the main cotton-producing area, which created a labour bottleneck with the principal cash-crop at harvest.

During field visits, farmers identified a number of plots where they had planted new varieties at the field level for the first time. If a new variety was obtained from the market, or another fairly anonymous source, farmers tended to manage it in the same way as similar varieties already in their possession. If, on the other hand, the variety was obtained from a friend or relative, the seeds often came with specific planting instructions, or as part of a management 'package' involving the variety and specific management practices or use of specific niches. The reported spread of a new indigenous production 'package' for growing rice in the water-retaining depressions of the hardpan plateaux (a niche used by farmers in other locations, but which had not been exploited by farmers in this particular area) consisted of both a new rice variety and management information (Moseley, 1993). Nor do farmers hesitate to combine new materials and ideas to form their own novel production 'packages' for experimentation, as was the case in one location where farmers were trying a new upland rice variety obtained from a neighbour, in combination with the use of their mechanical planter—a production technique passed on by a close friend in another village.

In addition to varietal trials, farmers also actively conduct experiments involving a number of other related aspects of their cropping systems. These additional experiments include major adjustments in the timing of sowing different crops, seeding densities and plant spacing, as well as numerous intercropping arrangements. Many of these adjustments have been undertaken in response to the regional downturn in rainfall. Farmers in the Far North Portfolio area, for example, were forced to abandon their previous practice of planting parallel rows of peanuts down each ridge. The initial response was simply to reduce the number of rows on each ridge from two to one, effectively halving their plant density. However, farmers in one village had developed a new planting method using a zig-zag pattern, alternating sides of the ridge with each seed pocket, which allowed them to maintain a higher plant density without the risk of inducing competition for the scarce soil moisture. By the second year of experimentation with this technique, the practice had spread to most farmers in the village.

Experimentation with different intercropping associations is also a common theme. Some of the more speculative efforts observed involved cash crops and were almost exclusively carried out in test plots located within farmers' home gardens—the closest analogy to an indigenous laboratory. One farmer in the South-west Portfolio area was running a series of intercropping trials that involved different spatial and temporal arrangements of bananas, tobacco and manioc. Others were experimenting with entirely new crops, such as pineapples and various market vegetables.

Agroforestry systems

In addition to field and garden crops, experimentation with agroforestry associations is another area of high activity among farmers. Across the entire OHVN zone farmers have begun to experiment with producing and planting a wide range of tree species in their main agricultural fields and home gardens. With encouragement from the Farming Systems Research Team, farmers in one village of the Far North were beginning to experiment with planting *A. albida* in their agricultural fields to help combat declining soil fertility levels, as well as to gain an additional cash crop (in this case, seed pods as fodder). In the Southern Portfolio areas, farmers were independently experimenting with planting *néré*, *karité* and other valued species directly into their main production fields. Individuals with grafting experience have developed their own varieties of mangoes and other fruit species, both for their own personal use and for sale.

Soil conservation and water control

Out of necessity, farmers in many areas were also actively experimenting with a range of technologies to improve soil fertility, and different methods of water control. Farmers in the Far North Portfolio area were experimenting with residue retention in their inner fields as a means of increasing soil fertility and reducing wind erosion during the dry season.[17] Elsewhere, farmers were trying several different types of physical control measures to help control the movement of runoff from their fields and lessen erosion problems. Examples of these different measures include the use of hand-dug canals, grass-lined waterways and vegetative barriers, and re-learning how to use rock barriers as a means of filling and controlling the spread of existing gullies.

Summary

In many respects, farmers' knowledge, communication pathways and ability to innovate represent the real life-blood of local adaptation and survival. Farmers possess detailed environmental knowledge, as well as knowledge of many important physical and biological processes, which they apply in carrying out the complex agricultural performances and management of diverse household production systems described earlier. On an individual basis, knowledge is acquired largely through personal experiences and overlapping communication pathways, both of which are strongly influenced by a number of social factors, including gender, age, ethnicity and family ties, amongst others. Farmers' creativity is not only a vital force in enabling each individual to meet her or his immediate personal and household needs but, as will be covered in Chapter 6, farmers' innovativeness has served as the major force in the adaptation and advancement of local production systems over the past 30 years.

5 The organization and actions of the formal agricultural research and extension system

As illustrated in the preceding chapters, farmers' knowledge and creativity have enabled them to make a constant stream of adjustments in their management of some very challenging production environments. Viewed over time, farmers' creative input has served as one of the major 'sources' of innovation in the evolution of local management systems, as will be highlighted in Chapter 6. Yet technological change does not take place within a vacuum. Without first establishing a comparative basis from which to interpret various contributions to change, it would be difficult, if not impossible, to fully appreciate the significance of farmers' role in the change process, to say nothing of the important role that their indigenous capacities could play in making future changes. This chapter, therefore, draws out those elements of the larger institutional milieux that have helped to define the range of options and operational processes, or 'technological space,' with which farmers have interacted and from which they have drawn ideas and materials. The chapter first introduces and then reviews those organizations that, regardless of their motivation, are separately and/or jointly dedicated to bringing about change in farmers' management practices.

From the outset it should be made clear that within the OHVN, formal agricultural development activities have been, and continue to be, dominated by the governmental research and extension programmes serving the area. Given this situation, the chapter focuses primarily upon these governmental efforts, their objectives and methods, and the outcomes generated by the tremendous investments that have been made in their operation over the past several decades. In undertaking such a review it is unavoidable that a number of critical observations be raised. The purpose of this analysis, however, is not to evaluate the performance of these organizations *per se*; this has been done elsewhere (e.g. Bingen *et al.*, 1993; McCorkle *et al.*, 1993; Kingsbury *et al.*, 1994; Simpson, 1995). Rather, the current assessment is intended to provide a brief but clear description of the range and nature of technological options that have either provided farmers with 'windows of opportunity' to adopt, or challenged them to adapt or generate, alternative technologies through their own efforts.

Formal extension and research efforts in the OHVN

Thus far, reference to the OHVN has generally been confined to its geographical boundaries, the OHVN zone of operation. Yet, as the principal development organization serving farmers in south-western Mali, the OHVN embodies other dimensions that are of critical importance in better understanding the dynamics of agricultural change in the area. Structurally, the organizational framework of the OHVN reflects an approach to rural development that has emerged from a unique blend of political/philosophical antecedents, patterns of financing and development objectives which continues to influence its orientation and range of activities (see Bingen and Simpson, 1997; Diallo, 1990; Jones, 1976; Simpson, 1995). Functionally, the OHVN and its major institutional partner, the Département de Recherche sur les Systèmes de Production Rurale, Volet OHV (DRSPR/OHV),[1] employ two of the more methodologically and operationally robust approaches to agricultural extension and research that have emerged over the past 20 years, namely the Training and Visit system of extension (T&V), and Farming Systems Research (FSR). Following a brief historical overview, this first section examines how these organizations, charged with the primary responsibility of stimulating agricultural change within the OHVN, have defined and discharged their tasks.

The genesis of the Opération Haute Vallée (OHV)

The agronomic potential of the Haute Vallée area began to receive serious attention in the early 1960s under Mali's first independent government (see Leynaud, 1962). By 1965, the French government had begun financing a development project in the Haute Vallée area, south of Bamako, which focused on the promotion of several commercial crops, shifting from rice and vegetable production to cotton to peanuts to tobacco (Steedman *et al.*, 1976). Despite the high hopes of the first regime for a village-led agricultural revolution, the decade of the 1960s generally witnessed a period of widespread stagnation in the agricultural sector. The one resounding success story during this period was that of the CFDT (Compagnie Française pour le Développement des Fibres Textiles), which, through the monopolistic powers it had been granted in the promotion and marketing of cotton, was able to stimulate a five-fold increase in cotton production.

Drawing lessons from the CFDT's success, and other development efforts, national planners began crafting the template for a new national agricultural development strategy (see Appendix B). Within four years of the 1968 *coup d'état*, the Second National Development Plan was unveiled, featuring a series of Opérations de Développement Rural (ODRs)—semi-autonomous, foreign-financed, development structures, responsible for the administration of commodity-based agricultural services in the different agro-ecological areas of the country (see CNPER, 1972). The ODRs were viewed as the primary vehicle for achieving the Malian Government's economic and rural development objectives and were given *carte blanche* to 'reorganize, co-ordinate and rationally utilize all of the means available to enable them to obtain their fixed objectives' (Min. de la Production, n.d.). In the Haute Vallée area, development efforts were reorganized as the Opération Haute Vallée (OHV).

From the outset, the OHV was an atypical product of the ODR formula. Unlike the other ODRs that were established with a clear commodity focus (e.g. cotton, peanuts, rice), the OHV lacked such a narrowly defined mandate, and attempted to embrace a much broader range of activities in its programming, including the provision of: agricultural extension, rural health care, co-operative formation, road construction, and functional literacy training, in addition to the promotion of a diversified range of livestock and cropping enterprises (Diallo, 1990).[2]

In 1978, the United States Agency for International Development (USAID) assumed financing of development activities in the OHV, and over the next 20 years would invest nearly $US 50million in helping to expand the portfolio of services offered to rural households within the OHV Zone (USAID, 1978; 1988; 1993).

The OHVN extension programme[3]

During the 1970s and 1980s, international financial shocks, followed by national policy reforms, led to a number of major changes in the OHVN's focus and geographic range of coverage. Declines in the world market prices of peanuts and cotton led to the collapse of the Opération Arachide et Cultures Vivrères (OACV), and forced the Compagnie Malienne pour le Développement des Textiles (CMDT) (formerly CFDT until 1974) to scale back its activities and concentrate on those areas of highest potential. Many of the areas vacated by the OACV and CMDT were transferred to the OHVN, which led to a tripling of the OHVN's geographic range of coverage, and a doubling of its client population. More importantly, these changes led to the introduction of a large, new, semi-arid agro-ecological zone for which the OHVN had no established package of technological recommendations.

Soon after the expansion of the OHVN's responsibilities, national policy reforms, based upon structural adjustment recommendations provided by the World Bank, led to a general restructuring of the ODR system (see SATEC, 1985). As a result the OHVN was required to shift a number of activities (credit, input supply and marketing) to the private sector, as well as to transfer 'greater responsibilities' for contracting these services to the growing number of village associations (Associations villageoise) (AVs). The reforms also forced the OHVN to reduce its number of field staff by nearly half (OHVN, 1988), and to turn increasingly towards a T&V style of extension to reach its clientele (see Appendix C).

The Training and Visit (T&V) Approach Following the basic strategies of the T&V approach (see Benor *et al.*, 1984), the OHVN has come to rely upon a three-tiered, cascade system of organization (headquarters, *secteur, sous-secteur*) to deliver technical messages to farmers in each of its ten administrative *secteurs*. Subject Matter Specialists (SMS) serve as *liaisons* with the different research programmes, and provide technical support and regular in-service training for field agents, *chefs sous-secteur* (CSS). The CSS in turn pass technical messages on to farmers through a regular schedule of bi-weekly visits with the all-male village extension groups (*groupement de vulgarisation*) (GV). Under this system, the CSS work most closely with contact farmers, or *paysans de contact*, the majority of whom manage small demonstration plots to showcase technologies promoted by the OHVN technical programme (OHVN, 1992c). In addition to its regular extension activities, the OHVN also manages a women's programme and helps co-ordinate annual farmer field days (*journées agricoles*), a limited number of farmer-to-farmer exchanges and radio broadcasts. By the early 1990s it was estimated that through the group meetings, field days and other contacts, over 26,000 farmers, or representatives of nearly one-quarter of the 40,000 households in the zone, were being contacted annually (OHVN, 1992b).

Technical content and orientation of the extension programme The content of the OHVN extension programme is structured around a number of technical themes (*thèmes techniques*) (see Table 5.1). Each of the themes, which can be roughly divided into four areas of emphasis (crop production, animal husbandry, natural-resource management, and women's activities), are summarized by a series of individual technical sheets (*fiches techniques*) that serve as the basis for discussions between the CSS and farmers.

Overall, the core of the OHVN crop- and animal-production themes reflect a package of well-known agricultural technologies, or *thèmes classique*, that have been the mainstay of agricultural development programmes throughout West Africa for many years.[4] The basic technical package emphasizes the combined

Table 5.1 ***The OHVN thèmes techniques**** (*Source*: OHVN, 1992a)

1a. Promotion of improved cotton varieties	9. Use of improved fallow
1b. Promotion of improved cereal varieties	10. Use of compost
2. Use of animal traction with improved cultural practices	11. Animal nutrition and health
3. Use of animal traction with improved varieties	12. Improved stables
4. Improved cultural practices	13. Animal husbandry—women
5. Understanding and use of fertilizer	14. Use of pesticides
6. Soil and water conservation	15. Use of herbicides
7. Market gardening	16. Field calculations
8. Animal fattening	17. Improved animal corrals

* Since this research was conducted four additional *thèmes* have been introduced: living fences; fire breaks; village tree-nurseries; and seed treatment (MDRE, 1997). These *thèmes*, however, were not part of the extension package during the period under research.

use of animal traction, standardized cultural practices, improved seed varieties and inorganic inputs (fertilizer, insecticide, herbicide). In response to recent donor concerns over sustainable agriculture, natural resource management and gender issues, the OHVN has begun to introduce supplementary recommendations covering these issues as well.

Affiliated programmes In addition to the OHVN extension programme, there are a number of international, governmental, non-governmental and private sector organizations operating within the zone that also provide farmers with information and materials.[5] Several of these are directly affiliated with the OHVN, and have established working agreements to carry out their activities in limited collaboration with the OHVN extension programme. The five principal programmes associated with the OHVN technical services are: the FAO Seed Multiplication Programme, which promotes decentralized seed multiplication through a network of villages as a means of improving farmers' access to new crop varieties; the pilot Agrométéologie programme, which relies upon 30 years of locally collected rainfall and yield data, a decentralized system of rainfall monitoring, and radio broadcasts to assist farmers in making improved decisions on the timing of their management operations; the United States Peace Corps, which posts volunteer 'teams' in the OHVN zone composed of specialists in education, water, agriculture, natural resource management and small enterprise development; the Projet Agro-Ecologie (PAE), a German-financed project that manages a series of farmer-assisted trials in expanding the use of soil- and tree-conservation techniques, and has provided in-service training for some *chefs sous-secteurs* in the use of GRAAP (Groupe de Recherche d'Action et d'Appui à l'Auto-Promotion) community resource management techniques; and finally, the Institut d'Economie Rurale's (IER) Département de Recherche sur les Systèmes de Production Rurale, Volet OHV (DRSPR/OHV), which serves as the principal source of new technical information and training for the OHVN extension programme.

The DRSPR/OHV Farming Systems Research Programme

Established in 1985, through USAID financing,[6] the DRSPR/OHV is charged with the primary responsibilities of developing agricultural technologies 'relevant to farmers' needs and circumstances', and of promoting the effective transfer of these technologies through improved linkages within the research system, and between research, extension and farmers in the OHVN zone (USAID, 1984:1) (see Appendix D). Since its inception, the DRSPR/OHV has conceived of and carried out a full programme of annual on-farm research and *pre-vulgarisation* (pre-extension) trials, in addition to conducting a number of base-line household economic and sociological studies (e.g. DRSPR/OHV, 1993c). These activities are managed through a network of research villages that are situated in four distinct geographic areas, or recommendation domains, identified by the DRSPR/OHV as being composed of households facing similar production conditions (e.g. soil and climate, among other factors) (SECID, 1987). Using criteria that are heavily oriented towards farmers' levels of livestock and equipment ownership (DRSPR/OHV, 1990b), researchers select and negotiate annual contracts with farmers in each research village to participate in the on-farm trials and household economic studies. Technologies that perform well in the research trials are moved to a one- or two-year phase of *pre-vulgarisation*, on-farm testing. Depending upon the results, a new technical recommendation and in-service training programme may be prepared for the OHVN field staff. By 1993 the DRSPR/OHV had prepared a farm household typology (see Table 5.4) and 16 *fiches techniques* (see Table 5.2) in support of the technical themes of the extension programme (DRSPR/OHV, 1993b).

Table 5.2 ***Fiches techniques* developed by the DRSPR/OHV** (*Source*: DRSPR/OHV, 1993b)

• Improved animal corrals	• High protein flour (niébé and cereal)
• Improved decortication of cereal grain	• Substitution of soy in making *Soumbala*
• Assisted natural reforestation	• Use of living fences
• Tree planting techniques	• Village wood-lots
• Use of micro-catchments for tree planting	• Reinforcement of banks along waterways
• Rock barriers	• Rock barrages
• Rock walls	• Rock lines
• Use of tied-ridges	• Tree wind-breaks

At the end of each year the results from the research and *pre-vulgarisation* trials are written up and reviewed in meetings with the other technical units, and in joint planning sessions held between DRSPR/OHV researchers and OHVN staff. Farmers' input on the research is solicited through annual, end-of-season, village evaluations held in each of the research villages (e.g. DRSPR/OHV, 1992b; 1993d). These evaluation sessions are open to both participating and non-participating farmers, who can provide 'feedback' on the technologies being tested, as well as make additional suggestions on new areas of research.

Prioritization and the research agenda The DRSPR/OHV has attempted to identify farmers' principal production constraints and use these in establishing its research agenda. Data obtained from two surveys (conducted in 1988 and 1990) have been translated by researchers into a series of research themes, and then prioritized on a three-tiered scale (see Table 5.3) (DRSPR/OHV, 1990b). Annual research trials are then designed to respond to these themes, e.g. running herbicide trials to address farmers' concerns over weeds and weeding labour.

Resource characterization To help target its own research efforts, as well as to assist the OHVN in directing its technical recommendations to the appropriate

Table 5.3 ***Farmers' perceived constraints and the DRSPR/OHV prioritization of research themes*** (*Sources*: DRSPR/OHV, 1990b; 1993c)

Farmers' constraints	Percentage of farmers reporting constraint	Research themes	Priority level
Weeds	49	Labour for weeding	1
Rainfall	39	Soil fertility and fertilizers	
Lack of fertilizer	22	Intercropping associations	
Soil infertility	19	Animal husbandry	
Lack of inputs	18	Disease, insects and birds	2
Birds	11	Planting dates	
Lack of draught animals	10	Crop storage and human nutrition	
Nematodes	7	Animal traction	
Insects	5	Income generation for women	
Labour shortage and varieties	4	Rotations and land management	3
		Soil preparation	
Disease and wind	2	Seeding	
		Plant densities	
		Labour at harvest	

target areas and groups of farmers, the research programme has constructed characterizations of both the physical resources and the economic status of households within the OHVN zone. In addition to the definition of four recommendation domains used in the research programme, the DRSPR/OHV has also constructed a three-part household typology (Yeboah *et al.*, 1991; DRSPR/OHV, 1992c), based largely upon the criteria of livestock and equipment ownership (see Table 5.4).[7]

Table 5.4 ***The DRSPR/OHV household typology*** (*Source*: DRSPR/OHV, 1993a)

Group level	Group descriptions		Percentage of population
1	UBT*	≥ 32	3.5
	Draught animals	≥ 6	
	Ploughs	≥ 2	
	Other equipment	≥ 1	
2	UBT	4–32	20.5
	Draught animals	≥ 3, but < 5	
	Ploughs	≤ 2	
3	UBT	≤ 4	76.0
	Draught animals	≤ 2	
	Ploughs	≤ 1	

* In this unit of measure, the *Unité Bétail Tropical* (UBT), one cow is equal to 0.7 UBT, and one sheep or goat is equal to 0.12 UBT (DRSPR/OHV 1993a).

Regional and private-sector research efforts

In addition to the DRSPR/OHV, other governmental research units and the regional ICRISAT programme, there are several other minor research programmes active in the OHVN zone. The major input supply companies, Comadis and Ciba-Geigy, separately and in collaboration with one or more IER research units, conduct their own research trials within the zone. These on-farm trials typically emphasize the use of fertilizers, pesticides/herbicides, varieties and equipment, and serve as both validation research and farmer-managed demonstrations of the companies' products. Such trials have generally been confined to the more productive environments of the southern OHVN zone, and within this area, to those farmers involved in cash crop production of cotton, maize and, to a lesser extent, sorghum. In all, during the 1991–92 agricultural campaign, the OHVN estimated that over 170 individual field trials were carried out in the zone by the various research programmes (OHVN, 1992a).

Contributions of the extension and research programmes

The remainder of this chapter focuses upon identifying the major outcomes and principal trends resulting from the combined and separate efforts of the research and extension programmes. The major features of the physical production environment, household resources, and farmers' needs for adaptive and highly diversified managerial responses, discussed in previous chapters (2, 3 and 4), are used as points of reference in assessing the inherent relevance of technical themes being investigated and promoted, as well as the approaches being utilized, by the two programmes.

The OHVN's technical programme

In considering the OHVN technical programme, three areas in particular deserve close attention: the appropriateness and coherence of the technical messages being offered to farmers; the ability of the extension programme to target these messages, or provide guidelines for their use; and the general flow of information and experiences within the programme, and between the programme and its partners, including farmers.

Appropriateness and technical coherence Despite the projected image of the popularity of the OHVN's technical programme, farmers, field agents and other observers have for some time identified a number of problems with the basic content and orientation of the programme.[8] In response to survey questions, nearly 70 per cent of the OHVN field agents noted that either economic constraints or perceived technical shortcomings of the primary extension package keep adoption rates low. These observations are confirmed by (the few) assessments conducted on adoption levels of specific technologies, which for improved sorghum varieties extended are estimated between 3 and 9 per cent (DRSPR/OHV, 1992d; OHVN, 1992a; cf. Matlon, 1990), and below 20 per cent for the maize production package as a whole. Economic analyses on individual technologies, e.g. the use of insecticides and ultra-low volume sprayers (Jago *et al.*, 1993), inorganic fertilizers in cereal production (DRSPR/OHV, 1988a; King, 1986;), use of mechanical seeders (Jaeger, 1986) and the *parc ameliorė* (McCorkle *et al.*, 1993), show that these technologies have limited applicability and are generally uneconomic for small-scale farmers in the drier areas. Despite these and other indications, however, the core package of technical themes has continued to be strongly promoted throughout the OHVN zone.

Other aspects of the OHVN extension programme reflect serious internal contradictions. Some of the new themes on natural-resource management (soil and organic matter conservation), for example, directly contradict the older *thèmes classique* that are still being promoted. One of the major emphases in the OHVN technical programme has been in encouraging farmers to increase their levels of mechanization, including the use of mechanical seeders. Farmers attempting to use the available seeders, however, are obliged to flat plough their fields, as well as to remove most of the crop stubble in order to keep the seeders from jamming and skipping. In addition to the increased time, labour and soil erosion potential associated with flat ploughing, the removal of crop residue further reduces a field's soil- and water-conservation capacity, as well as removing a major source of soil organic matter. In sum, the standing *thèmes classiques*, which tend to be based on a high-input model, are often diametrically opposed to the basic intents and specific practices contained in the newer natural resource management themes.

Targeting Despite the identification of recommendation domains and the development of a household typology by the DRSPR/OHV, there is no evidence that either of these tools have ever been used by the extension programme in attempting to target its technical recommendations to specific farm-level conditions; a shortcoming that the organization itself has recognized, which, in an internal draft report, observes that 'one can remark that in general the *thèmes* are disseminated without consideration to where they are pertinent (or expressed as needs by farmers)' (Anon./OHVN, 1992). For example, the promotion of inorganic fertilizer use, one of the most widely popularized technical *thèmes*, has been extended to GVs across the entire zone, including those in the Near and Far North Portfolio Areas, yet less than 5 per cent of the farmers in the northern areas have adopted the technology. For farmers in the Far North area, neither the OHVN nor private banks will provide credit for the purchase of chemical inputs, expressly because of the high risks involved. Even if the programme was interested in selectively extending technologies, as currently written, most of the *fiches techniques* do not contain the type of information that would allow them to be effectively targeted to specific areas or conditions (e.g. the *fiches* on varieties often do not contain information on preferred soil types or rainfall requirements).

With respect to those technical *thèmes* that do enjoy widespread usage (e.g. animal traction), the OHVN has needlessly limited itself in the scope of its intervention. For example, the technical recommendations on animal traction (AT) have focused exclusively on the use of bullocks, whereas farmers rely upon both horses and donkeys, in addition to bullocks, in performing field operations (e.g. DRSPR/OHV, 1991a). In fact, in some areas of the Far North Portfolio area, the importance of these alternative forms of AT outweighs that of bullocks. Yet after nearly 30 years of effort, neither the research nor extension service has broadened its conceptualization of AT to include other sources of draught power in promoting its potential.[9] Another example of this type of self-imposed limitation can be seen in the extension messages issued on rocklines, which have been promoted solely on their merits of conserving soil, and not moisture, despite the fact that in water-deficient environments moisture conservation provides the larger and more immediate yield response (e.g. rocklines can increase crop yields from 35 to 55 per cent (Matlon, 1990), and up to 80 per cent in drought years (Reij, 1991).

The flow of information The effectiveness of the T&V approach is predicated on the continued flow of new and relevant technological innovations from the research system to farmers, as well as the return flow of feedback from farmers and field agents to research. Although the OHVN extension programme has made great strides in streamlining its activities and monitoring the actions of its field agents, the programme has continued to experience a number of difficulties in the preparation and distribution of new technical materials. Six years after installation, SMS, who are charged with the preparation of new *fiches techniques*, have developed little new extension material (the exception being a collection of recently assembled *fiches* on vegetable gardening). Apart from its ties with the DRSPR/OHV research programme, there is no evidence that extension has made any use of other sources of information.

Included in this self-imposed isolation from alternative information sources is the lack of any appreciation of farmers' extensive agricultural knowledge. On the contrary, farmers are widely perceived as being one of the major obstacles preventing the programme from reaching its objectives,[10] and in dire need of 'sensitization' to more 'rational' practices (i.e. those offered through the *thèmes techniques*). Interestingly, the recent introduction of technical material on gardening, use of living fences, agroforestry, conservation tillage and grafting techniques can be seen as an effort by the extension programme to catch up with farmers' current practices.

To the extent that field agents are aware of farmers' concerns, interests and activities, there is little evidence that they either communicate their knowledge to researchers, or exchange experiences among themselves in order to contribute to the development of the programme. Little has changed since the observation in a 1985 evaluation of the programme, that 'such innovation [on the part of field agents] is positively discouraged' (RONCO, 1985:98) (cf. Kagbo, 1986; Lebeau, 1986; USAID, 1982). In their questionnaire responses, over 45 per cent of the field staff reported knowledge of a local variety or indigenous practice that, in their judgement, was superior to the technical recommendations being extended to farmers. Yet fewer than 7 per cent reported ever passing their observations on to others in the extension service, or to researchers.

The end result of the various problems with relevance, as well as the indiscriminate and poor flow of information, is clearly visible at the field level. Indicative of the static and marginally relevant nature of the technical information being promoted by the extension programme, a majority of the CSS report considerable difficulty in convening GV meetings. When GV meetings are held, attendance averages six farmers, less than half of the established upper limit of 15 persons (Anon./OHVN, 1992). Moreover, the same farmers do not attend on a regular basis; GV membership tends to 'roll' as members attend a few meetings and then drop out. The existing *fiches techniques* are neither widely available to the CSS or sector offices, nor are they directly accessible to farmers; virtually none of the technical recommendations have been compiled and translated into Bamanan for widespread distribution through the functional literacy programme (which was established to enhance farmers' access to new information). Supplemental activities, such as farmer-to-farmer visits, *journées agricoles* and radio broadcasts, have been woefully under-utilized, or subverted for other purposes (see Simpson, 1995). The women's programme in particular has suffered from a major under-investment.[11] Overall, the inflexible nature of the T&V system neither provides field agents with the opportunity to become involved in other programmes, nor allows other programmes, including those with which the OHVN has established working agreements, to contribute to the OHVN extension objectives.

The DRSPR/OHV Research Programme

In its attempt to contribute to agrarian change in the OHVN zone, the DRSPR/OHV has concentrated its efforts in three general areas that warrant consideration: the development and use of various 'tools' to help the research and extension programmes target their efforts; the creation and strengthening of a number of communication linkages between the DRSPR/OHV and extension, farmers and other research and development organizations; and the generation of new technologies with the potential of improving the productivity and economic well-being of farm families residing within the OHVN zone.

Use of research 'tools' On the whole, the research programme's translation and prioritization of farmers' production problems into specific field trials reflects some of the same bias towards high input technologies that is evident in the extension programme's *thèmes classiques*. For example, one of farmers' chief concerns, rainfall, is addressed by research only in terms of 'planting dates', and is relegated to second-order status in the research programme's three-tier hierarchy (DRSPR/OHV, 1990b), where it has received almost no research attention.[12] Other issues that were low priority for farmers, such as disease and insect pests, have been given higher priority rankings by researchers, and pursued through a high external input perspective, while *thèmes* such as poor intercropping yields and animal husbandry, which were not identified as concerns at all by farmers, have been given first priority within the research programme (see Table 5.3).

As would be expected, the first priority themes have clearly dominated the agenda of the field trials. This emphasis, however, masks another form of bias that has lead to the majority of cropping-system trials to focus on the use of purchased inputs—various fertilizers, herbicides and insecticides (see Table 5.5).

Table 5.5 ***Variables examined by the DRSPR/OHV Research Programme (1990–91; 1991–92; 1992–93)*** (*Source*: DRSPR/OHV, 1991a; 1992a; 1993a)

Variable tested (priority level)*	1990–91 (# of trials)	1991–92 (# of trials)	1992–93 (# of trials)	Total
Fertilizer (1)	8	5	7	20
PNT** (1)	6	2	8	16
Animal husbandry (1)	2	3	3	8
Herbicide (1)	2	1	3	6
Intercropping (1)	2	2	2	6
Manure (1)	2	0	0	2
Mechanical weeding (1)	0	1	0	1
Insecticide (2)	2	2	0	4
Sources of revenue (2)	1	1	0	2
Crop storage and consumption (2)	1	0	0	1
Timing of sowing (2)	0	1	0	1
Variety (3)	4	0	0	4
Fallow (3)	0	1	1	2

* Each DRSPR/OHV research trial commonly examines several variables, e.g. the simultaneous comparison of varieties, various fertilizers, and manual versus chemical weeding. The data summarized in Table 5.5 attempt to include each of these individual variables that were tested in field trials conducted during the three research campaigns examined.

** PNT = Phosphate Naturel de Tilemsi

Such an emphasis is neither consistent with the economic abilities of the majority of households in the zone, nor with the general environmental conditions of the zone, e.g., highly variable rainfall.

Surprisingly, as was the case with the extension programme, there is no evidence in the research programme's annual reports, planning documents or informant observations indicating that the household typologies or recommendation domains developed by the DRSPR/OHV have been used in targeting research to specific social-economic groups or physical conditions (e.g. DRSPR/OHV, 1988a; 1989; 1990a; 1990b; 1991a; 1992a). As noted, one result has been that, on economic grounds alone, the major thrust of the research programme in promoting the higher use of purchased inputs is inappropriate for the level of financial resources controlled by most households in the OHVN zone (McCorkle *et al.*, 1993). The preponderance of research trials, shown in Table 5.5, revolve around the use of specialized equipment and/or purchased inputs. Yet the DRSPR/OHV's own research indicates that less than 25 per cent of the households in the zone are likely to have the financial means and equipment necessary to take advantage of these types of technologies (DRSPR/OHV, 1993a). Furthermore, in order for poorer households to make use of these technologies, they would have to increase significantly their exposure to financial risk, in an environment when erratic rainfall and acute pest outbreaks often result in poor harvests—a level of risk that, in many cases, not even the banks are willing to support.

At a more fundamental level, there has been a general failure within the research programme to recognize and respond to the inherently diverse and variable qualities of basic resources that farmers depend upon in different geographic areas. On the one hand, the research results from years with poor rainfall tend to be treated as aberrations when testing new technologies, instead of as an endemic feature of the environment to which new technologies must be responsive. Yet even those efforts geared towards helping farmers make better use of the available rainfall oversimplify the challenges to a point where they lose relevance. The recommendations on planting dates, for example, issued through the pilot Agrométéologie programme, are based upon the use of decentralized rainfall monitoring at the village level. However, due to the highly disparate nature of local rainfall, farmers' fields, which can be several kilometres distant from the centrally placed rain gauges, often experience very different rainfall patterns. When combined with differences in topography and the moisture-holding capacity of different soil types, such a simplistic approach is hardly adequate for highly specific situations.

A similar observation can be made with regards to differences in soil type. Despite the fact that most agronomic decisions made by farmers include reference to the characteristics of the soil type involved, the research programme has no apparent strategy for characterizing, incorporating, or using soil heterogeneity in its programming. A system, such as that developed by ICRISAT in using a soil 'toposequence' approach in running its varietal and intercropping trials (ICRISAT, 1979), has not been adopted by the current DRSPR/OHV programme.[13] In many ways, the apparent separation between the research themes on cropping systems improvement and natural resource management (indicative of an underlying tension between low-input and high-input approaches) is linked to the noted gaps between farmers' economic and physical realities and the priorities, tools and methods employed by the research programme.

As pointed out by McCorkle *et al.* (1993), the near complete lack of women's involvement in the research programme is another way in which the DRSPR/

DHV programme has needlessly limited its potential impact. Women participate in the majority of agriculture activities throughout the OHVN zone, and provide a major share of the agricultural labour. Yet women have neither featured prominently in the major baseline surveys, nor have they participated in the majority of research and *pre-vulgarisation* trials (as with the OHVN extension programme, the DRSPR/OHV has attempted to address women's concerns through a discrete set of studies, which, once separated from the main set of activities, have tended to be under-funded). This fact, that from the outset, one half of the zone's farmers are excluded from participating in the primary research and development activities, has serious implications for the impact of the research programme as a whole.

In sum, by failing to appreciate and respond to the inherently diverse and variable qualities of the physical environment, or the need for farmers to utilize equally adaptive management practices, the formal system has prevented itself from acquiring an informed understanding of what it takes to be a good and successful farmer under the prevailing conditions. On the whole, farmer behaviour is not an issue that has been critically examined by the research programme. Although statistics are collected annually on a number of specific aspects of *what* farmers do, e.g. levels of equipment usage, areas planted to specific crops, as well as more detailed studies on household agricultural revenues and expenditures (e.g. DRSPR/OHV, 1992a; 1992c; 1993a), virtually no data have been collected on *how* or *why* farmers engage in any of these specific practices. At best, there exists an implicit appreciation among individual researchers that farmers are rational and highly efficient maximizers of resources. Yet this appreciation falls far short of a systemic understanding of the basic need for farmers to develop and maintain a flexible outlook in their approach to agricultural production—especially in terms of the types of technologies that they require.

The flow and exchange of information One of the major objectives of the DRSPR/OHV programme has been to establish and improve the working linkages between the research unit and its numerous development partners (USAID, 1984). In the case of farmers, this functional relationship with DRSPR/OHV researchers appears to be minimal. The annually negotiated contracts between the research programme and farmers participating in the on-farm trials, while a necessary instrument for articulating each party's duties and responsibilities, has neither been constructed nor used as a foundation for establishing an interactive partnership between researchers and farmers. Farmers' involvement in the trials has been generally limited to the contribution of land and labour (e.g. DRSPR, 1990b; 1991c; 1992e). The rationale for undertaking each research trial is that of the researcher, leaving farmers virtually no 'voice' in setting the research agenda (see Norman, 1980). Much of the emphasis in strengthening farmer–research relationships appears to be placed on the recently established post-research feedback sessions, *evaluation paysanne*, which are now conducted annually in each research village (e.g., DRSPR/OHV, 1992b; DRSPR/OHV, 1993d). While these sessions are conducted in a highly efficient manner, the fact that they are carried out on only an annual basis prevents them from serving as a forum for active researcher–farmer collaboration along evolving research themes. In the few instances where farmers' knowledge and input have been incorporated into the research process, the results have been resoundingly positive. The *fiche technique* on using soy as a substitute for *néré* seeds in making soumbala is based upon farmer input (as well as that of the national nutrition lab), is one of the few *fiches* developed that involved original research (DRSPR/OHV, 1993a), and has been rightly hailed as an example of 'FSR/E at its holistic best' (McCorkle *et al.*, 1993). Interestingly, this example also represents one of the few research efforts into the minor theme on 'women's issues.'

In regard to the relationship between the DRSPR/OHV and its other institutional partners, the research programme and the OHVN extension service both report highly effective and amiable relations with one another. Yet when the substantive issue of technology generation is considered, the managers of both programmes admit that the exchange of relevant information is a major weakness impeding the success of agricultural development efforts in the zone (OHVN, 1992c; Sélingué, 1992; Teme *et al.*, 1993). With respect to the quality of the DRSPR/OHV's relationship with other ODRs and government research units, the experience appears to be mixed. On the one hand, at least 12 of the 16 new *fiches techniques* prepared for the OHVN extension service by the DRSPR/OHV originated from research conducted elsewhere. Yet not all inter-organizational linkages have been entirely beneficial for the research programme. It has been suggested that the evident bias in the on-farm testing programme towards high input technologies is the result of influences exerted by the more traditional, discipline-based, and centrally positioned research units within the IER.

The problems associated with the uneven quality of external relationships are compounded by the programme's generally weak use of literature and the existing knowledge base. Key informants indicated that a thorough review of the literature is generally not done prior to the design of each research trial (as can been seen in programme documents, which rarely offer a rationale for why a particular technology was selected for testing). Intercropping, tied-ridges, use of natural rock phosphate and various inputs, amongst others, are all themes that have been extensively researched in Mali, and elsewhere in the region, beginning as early as the 1970s. The cost of failing to review and make better use of the existing knowledge base is redundant, outdated and misguided research, which further postpones the creation of needed technical alternatives.

The generation of new technical recommendations As outlined in its mission statement, the *raison d'être* for the DRSPR/OHV research programme is the adaptation and generation of new technologies relevant to the conditions faced by farmers in the various parts of the OHVN zone (USAID, 1986). Although it is difficult to determine the exact impact of any research programme, and particularly so in the case of the DRSPR/OHV (which did not begin to monitor its progress until 1991), several indicators exist which hint at the programme's probable influence on agricultural change within the zone. Overall, it is difficult to ignore the fact that while the research programme has issued a range (albeit limited) of new technical recommendations to extension,[14] the results of the *prevulgarisation* trials for the programme's first six annual reports (1987–88 to 1992–93), those trials which are the final stage of verification before passing new technologies on to the extension system, show that none of the agricultural production technologies examined produced statistically significant results (DRSPR/OHV, 1988a; 1989; 1990a; 1991a; 1992a; 1993a). In essence, while the DRSPR/OHV research programme has adopted a limited number of technologies from other research programmes, it has not developed any viable production alternatives through its own research. In addition, for those technologies that have been compared to farmers' existing practices, there is no indication that when selecting 'traditional practices' for use as controls, those selected practices represent anything close to farmers' 'best practices'.

Summary

While attending a meeting of a farmer-designed and run on-farm research programme in a different part of the world, one of the participants rose and made the observation that 'so many good ideas came at us so fast [from the research system] that we began to expect that research would always provide for us', with one of the consequences being that farmers gradually 'lost [their] ability to carry out [their] own investigations'. While highly relevant and productive research programmes certainly have the potential of contributing to a decline in farmers' experimentation, there is little evidence to suggest that this has been the case within the OHVN. The preceding review of the historical and contemporary range of research and extension activities indicates that the dominant development organizations operating in the OHVN have had only a modest impact on shaping farmers' 'technological space'—the pool of technologies, materials and processes with which farmers interact and draw from in adapting and improving their production activities and general welfare. Despite the sustained financial support that research and extension activities have received since the 1970s, a number of features associated with their orientation, organization and modes of operation have largely prevented them from achieving greater successes. In the end, the resulting oversights and shortcomings of the formal development efforts have set the stage for farmers to have not only the opportunity, but also the need to develop technologies of their own.

6 The nature and roots of change in household production systems

Farmers' own innovations, and their adaptation and adoption of technologies developed and promoted by the formal research-extension system, amongst others, has led to a number of significant changes in the household production systems in the OHVN zone over the past several decades (see the section on site selection in Appendix A). Due to the multiplicity of 'sources' of innovation, as well as the many different forms that change can take, the challenge of attributing a specific change to a specific source is daunting. In addressing this challenge, the information presented in this chapter is drawn from data collected on farmers' observations over three general time periods: the general changes occurring since the 'time of their mothers and fathers' (roughly corresponding to the pre-drought period, prior to the late 1960s); the changes made in the recent past (the last three to five years); and any 'new' activities that farmers were attempting to introduce during the year in which the fieldwork was carried out (see Appendix A). The information presented on the longer-term, inter-generational changes reflects the generally normative, though sometimes dissenting, responses originating from the group interviews. In contrast, the identification of more recent changes was aided by more intimate, one-to-one discussions concerning specific technologies or management challenges, and on-the-spot descriptions of field activities and histories carried out with individuals and small groups of farmers.

The results of all village-level data were then compared with a thorough review of DRSPR/OHV research records and the *thémes techniques* issued by the OHVN extension service. Interviews with key field and administrative staff, supplementary research reports and secondary material, as well as a more general survey of the technical content of other projects and programmes operating in the OHVN zone, provided additional points of reference.

Though far from exhaustive, the following review helps to depict the general nature and balance of the contributions made by the principal 'sources' of innovation in the evolution of household production systems during the lifetime of the current generation of agricultural managers.

Cultural tools and practices

Certainly one of the most visible and widespread changes to occur in farmers' production systems since the 'time of their mothers and fathers' has been the spread of animal traction (AT) and use of the plough. First introduced into the OHVN area during colonial occupation in the early 1930s (Leynaud and Cisse, 1978), AT has since been aggressively promoted by various governmental and bilateral programmes and, more recently, non-governmental organizations and private sector-interests throughout the southern portions of the OHVN zone. According to farmers' estimates in different villages, between 40 and nearly 100 per cent of their farm land is currently cultivated using AT. In general, poorer households and women still rely most heavily upon manual cultivation, although consecutive bad years and unique circumstances can force even relatively wealthy households to divest of their capital assets and revert to manual cultivation—a pattern of investment and divestment found across much of the Sahel.[1] Although farmers using AT perceive a major increase in their ability to cultivate more land (as well as relief from the drudgery of manual cultivation), empirical studies in the OHVN zone, elsewhere in Mali and the region, show that the use of the plough has had little effect on the amount of land cultivated per individual worker; at most, an insignificant increase or, in some cases, even a slight decline (e.g. Adesina, 1992; Coulibaly, 1987; IER, 1978; Jaeger, 1986;).[2] Labour bottlenecks involving operations that are still largely performed manually, or for which superior mechanical options have yet to be introduced, such as thinning, transplanting and in-row weeding, continue to prevent major increases in the amount of land cultivated per worker. While the hiring of tractors for ploughing is on the rise, it is still limited to the wealthier producers in the Central and Southern Portfolio areas.

While farmers in the OHVN seem to have largely adopted AT implements without making major physical modifications in their design,[3] the cultural practices associated with their use have undergone significant adaptation to fit local conditions. Farmers' traditional mound culture, once common throughout sub-Saharan Africa (e.g. de Schlippe, 1956; Miracle, 1967; Knight, 1974; Warner, 1991),[4] has been largely replaced not by the practice of flat ploughing, the *théme classique* long-promoted by the extension service, but by the use of ridges, an indigenous adaptation created by making consecutive dead-furrows with the plough. As the formal research system is discovering, farmers' use of ridges is not only labour saving and effective in protecting soils against erosion, but is also more effective than flat ploughing in conserving soil moisture, the principal constraint on crop production in much of the OHVN zone (e.g. SRCVO, 1992; cf. Klaij and Hoogmoed, 1989). In the main cotton-producing locations of the Southern Portfolio areas, where the use of flat ploughing is more widespread, farmers have observed that this practice is also accompanied by a significant increase in soil erosion.

The increase in mechanized cultivation has neither replaced nor significantly altered the range of different *dabas*, the principal hand tool used by farmers in their cultivation, planting, thinning and weeding operations.[5] Yet the introduction of AT and the development of new ploughing techniques have had a secondary influence on the 'styles' of manual cultivation performed. Although the use of mounds is still found, albeit on a much reduced scale, farmers relying upon manual cultivation commonly mimic the ridges produced with a plough. As with mounds, ridges bury weeds and concentrate the fertility of the scant top soil in the seed bed. In both cases, runoff is slowed, which allows for greater infiltration of rainfall when compared to flat ploughing.[6] In agreement with the more recently issued extension messages on conservation tillage, farmers have typically aligned their ridges 'across the slope' to enhance the soil and water conservation potentials, although they often do not follow the changes in contour. In contrast to both old and new extension messages, farmers in areas with steeper slopes have developed a system of creating wide (across-the-slope) but shallow (down-slope) fields with the furrows oriented downhill, or on an angle, to facilitate the rapid evacuation of excess rainwater. This practice eliminates the risk that rainwater, trapped behind a ridge during the frequent torrential downpours, might break free and cause a complete washout of the hillside (a strategy also used by farmers in hilly areas of central and eastern Africa, as well as in areas of South America). Farmers using this practice reported less concern about the loss of soil on hillsides than the burial of young plants down-slope.

As a natural complement to the use of ridges, there has also been an increase in the use of row planting. With the exception of those crops that are still commonly planted by broadcasting seeds, such as fonio and rice, most cereal crops are now planted in rows, regardless of the method of land preparation. Even in highly integrated intercropping associations, such as women's mixing of peanuts, Bambara groundnuts, millet and okra or *dah*, one or more of the species are often planted in rows (which was not the case under the previous system of mound culture).

In comparison with the popularity of the multiculture and standard plough, the use of other implements, such as the harrows and mechanical seeders promoted by the extension service, is much lower. After widespread experimentation, farmers generally cite high costs and low utility as the major reasons why they have not made further use of these technologies; such sentiments are supported by economic analysis, which suggests that mechanized seeders are generally uneconomic for cereal farmers in the Sahel (Jaeger, 1986). In some cases, farmers have abandoned the use of certain pieces of equipment after an initial trial period. Such is the case with mechanical seeders, which now sit idle in many courtyards across the zone. These seeders, in order to operate effectively, require fields that are flat-ploughed and generally free of crop residues (as residues cause the seeders to skip and jam), conditions which, as noted in Chapter 5, are openly counter to farmers' traditional soil- and moisture-conservation practices, as well as many of the new natural resource management themes being promoted by the extension service.

Those farmers who have not abandoned ineffective technologies outright have had to undertake significant adjustments, and continue to experiment with a number of related aspects of their management practices. Farmers in the Near North Portfolio area, using mechanical seeders with their cereal crops, have had problems obtaining good crop stands because of the reduced number of seeds placed per pocket. Low germination rates, combined with insect and animal pest damage, force farmers to bring transplants in from other fields, which they have hand planted, in order to fill in the resulting 'gaps,' rather than the usual practice of thinning and transplanting within a single field. Farmers in some locations of the South-eastern Portfolio area have found that under their field conditions the plant spacing of cotton rendered by the mechanical seeder is too close, forcing them to thin each plot manually. In an attempt to get greater use out of their seeders, farmers in the main rice-growing areas have begun to experiment with using them to plant upland rice, a practice which is not part of the OHVN extension package, but which has been promoted in the large-scale irrigation schemes elsewhere in the country (e.g. Office du Niger; Opération Riz Segou and Opération Riz Mopti).

Land use and fertility management

The perceived decline in annual precipitation levels has led farmers across the zone to make numerous adjustments in their traditional patterns of land use. In the past, farmers, even in the northern-most areas, were able to produce limited amounts of rice in the seasonally inundated areas. Now, however, these sites are used for maize and sorghum production, and in some cases are considered prime land for millet, farmers' most drought-tolerant crop. In the South-eastern Portfolio area, Moseley (1993) found that the climatic shift has also led to the revaluation of local land types. Villages in this area, with an abundance of free-draining soils, ideal for cereal production under high rainfall conditions, found themselves in a situation of land deficit, with shortages of moisture-retentive soils necessary for sustaining production levels under a lower rainfall regime. As a result, farmers have not only had to adjust the types of crops grown on different sites, but also negotiate new land tenure arrangements with neighbouring communities who have a surplus of more suitable land types.

In addition to climatic change, farmers across the OHVN point to growing population pressures as another major contributor to declining soil-fertility levels. Farmers in the Central and South-eastern Portfolio areas made the additional observations that the increased use of the plough and long-term use of inorganic fertilizer (elements of intensification possibly linked to demographic changes) had also contributed significantly to their soil fertility problems. In response to growing land pressures, farmers in roughly half of the communities visited reported making substantial reductions in their fallow length when compared with the preceding generation (with farmers in several areas making the transition to continuous, or near continuous, cultivation of the greater part of their crop land). The majority of farmers also reported decreasing the length of their cropping cycle, due primarily to declining fertility levels and rising weed pressures. One of farmers' most problematic weeds, striga (*Striga spp.*), is noted to flourish in impoverished soils, which farmers in some areas have linked to long-term use of the plough and the associated reduction in organic matter through the removal of trees and mixing of infertile subsoils. Interestingly, farmers value the application of inorganic fertilizer least for restoring soil fertility and combating invasions of striga. Use of a legume rotation, fallow or, when possible, the addition of animal manure and household wastes were all considered more effective.[7]

The extension service's long-standing, blanket recommendations for inorganic fertilizer use has recently been augmented by additional recommendations on standardized manure and compost application, as well as fixed cropping rotations. While generally ignoring the overall recommendation 'package', farmers have widely adapted and increased their use of the individual components involved. For example, although farmers and researchers alike have long recognized the economic non-viability of blanket applications of inorganic fertilizer in the northern half of the OHVN (e.g. DRSPR/OHV, 1988a; King, 1986),[8] farmers do make selective use of inorganic fertilizer, even in the driest areas. The extension programme can also be credited with the increased manufacture and use of compost, as well as rocklines, both of which are traditional practices.

Although farmers across the OHVN claim that rock- and grass-lines are a traditional soil- and water-conservation practice, few have maintained active use of the technology, possibly due to the growing popularity of the plough, which makes negotiating around fixed obstacles difficult.[9] As a result of extension efforts, and the informal spread of the practice, farmers in a number of locations are re-adopting the use of rock barriers in their fields to stop the spread of gullies. The use of manure has also increased, in sync with the rise in livestock ownership and growing concerns of fertility management. Although the extension programme has promoted the increased use of manure, the recommended application rates generally exceed farmers' supplies and capacities to apply. A market for manure may have developed in some areas (McCorkle, pers. comm.), though in general the sale of manure is viewed as a fairly desperate measure, because it represents a sale of productive potential.[10]

As was noted in Chapter 4, farmers in different locations have been actively experimenting with several new management techniques—tree planting, residue retention, and a number of physical control measures—to help control water runoff, reduce erosion and maintain and improve soil fertility levels. With and without governmental assistance, farmers in all parts of the OHVN zone have been making increased use of movable corrals situated in their main production fields, the construction of large barrages in *bas fond* areas to create more rice land, compost pits, and other practices in the attempt to improve the production potential of their lands. While the use of each of these innovations is uneven, the clear trend is towards an increase in attention to land management issues by farmers.

Cropping systems

Although far less visible than the changes occurring in other areas of farmers' production systems, i.e. the substantive increase in the use of AT, local cropping systems have undergone the most extensive transformations since the preceding generation. Included are major changes in the crops and varieties used, rotations and related cultural practices—timing of sowing, seeding densities and plant spacing—to match reduced moisture levels, as well as the development of new intercropping arrangements and the addition of new crops to exploit emerging market opportunities. In the Far North Portfolio area, farmers have found that the decrease in rainfall has been accompanied by an increase in the variability of early rains, such that they can no longer safely plant peanuts prior to the onset of the main rains as they had in the past. Farmers now attempt to time their planting to follow, as closely as possible, the onset of the rains.[11] As described in Chapter 4, the lower rainfall has forced farmers in the northern areas to change how they plant certain crops. In an attempt to maintain a higher plant density, while at the same time avoiding excessive moisture stress due to competition, farmers have begun to plant peanuts in a zig-zag fashion on alternating sides of the ridge, after first shifting from a double- to single-row planting strategy. During the preceding generation, farmers in the Near North Portfolio area typically planted cowpeas as the first crop in their rotations, just as a field was taken out of fallow. Now, however, due to decreased rainfall, farmers prefer shorter season cereal crops that are less affected by the increased variability of the early rains. In the South-western Portfolio area farmers have had to abandon completely their occasional double-cropping practices. In response to the early periods of drought, farmers across the Southern Portfolio areas replaced much of their former (aquatic) rice production with maize, a crop strongly promoted by the extension service. Yet during more recent years with slightly higher rainfall, many of these maize fields have become waterlogged and produce poorly, if at all. As a result, many farmers have shifted back to the cultivation of varieties of upland rice that can tolerate higher moisture levels, with the areas planted to maize falling off sharply.

In adjusting to the new environmental conditions, farmers have also shown an interest in new sources of information. Nearly all farmers in the Central and Southern Portfolio areas of the zone expressed an appreciation for the information on planting dates broadcast on the radio by the Pilot Agrométéologie programme.[12] While none of the farmers had ever based their planting decisions solely on this information (i.e. planted when announcements were made), they did use it as an additional source of information to complement the many natural signs upon which they rely in determining when to begin sowing their crops in particular fields (see Box 4.1).[13] The temporal and spatial diversity of rainfall throughout the region (see Chapter 2) is such that individual fields, often located several kilometres from each other, receive entirely different patterns of rainfall which, when combined with the different moisture-holding capacity of various soil types, makes blanket planting recommendations irrelevant at best.

In recent years the DRSPR/OHV has placed greater emphasis on investigations into different spatial arrangements in cereal–legume intercropping associations. While this shift from a fixation on mono-cropping systems that dominated much of the 1980s more closely resembles farmers' current practices, the focus on developing standardized recommendations for plant densities and planting patterns misses entirely the point of farmers mixing and matching their intercropping practices to fit the conditions of specific fields (soil types, fertility status, moisture regimes and weed populations).[14]

New varieties and new crops

In preparing a first line of defence against the many production challenges—varying soil fertility and moisture levels, endemic weed pressures, disease, and insect and animal pest problems—farmers select and attempt to maintain a range of varieties from among the local stocks of genetic material. Over the past decade the physical and economic changes occurring in the OHVN have led to several notable trends in farmers' additions and deletions of varieties from their repertoire of planting material. The increased aridity, for example, has led to a general increase in the adoption of millet, sorghum and maize varieties with faster maturation periods and greater drought tolerance.[15] Most of the short season varieties introduced by the OHVN, however, have enjoyed only limited success with farmers. Poor taste and field performance in comparison with existing varieties are the principal reasons cited by farmers for not adopting the new varieties. As one researcher in the farming systems research programme (DRSPR/OHV) observed, 'there are hundreds of local varieties in the zone that are better adapted to local conditions'. In an attempt to mediate these problems, the FAO decentralized Seed Multiplication Programme has adopted the strategy of using only improved local varieties in their multiplication programme, with some success.[16]

Farmers continue to seek out and select varieties with specific characteristics to meet the constraints imposed by evolving conditions in their production environments. In the south, higher-yielding varieties of Asian rice (*Oryza sativa*) have largely replaced local varieties of African rice (*O. glaberrima*). Yet farmers in several areas have begun to re-acquire varieties of the local, red-hued African rice, which tend to be more tolerant of the prevailing conditions of low soil fertility and high levels of weed infestation. Farmers have also begun to widely adopt local varieties of sorghum, such as *seggatana* (*segga* is the Bamanan word for striga), that are resistant to striga.[17] In addition to weeds, bird predation is another severe production problem facing farmers. In some areas, the severity of bird predation is such that farmers have had to virtually abandon large-scale millet cultivation.[18] In these locations, the adoption of local varieties of millet with stiff, sharp bristles that protect the grain head from bird attacks is on the rise. At one time the research programme had been experimenting with an intercropping system using tall, traditional, bird-resistant varieties of sorghum as a 'guard crop' for dwarf millet. Farmers who had been taken to visit this demonstration plot were very interested in trying out the technology, but after failing in their attempts to obtain seeds of the dwarf millet variety over a two-year period, had given up.

Some longer-term changes can be observed in farmers' use of individual crops over the past several decades. Well before the creation of the OHVN, farmers reported that they had replaced their existing varieties of short staple cotton with new, long staple cultivars introduced in Mali by the French cotton parastatal, the Compagnie Française pour le Développement des Fibres Textiles. Today, cotton is no longer planted as part of the regular intercropping mixes, but is now grown exclusively as a commercial mono-crop. Across the Southern Portfolio Areas cotton has replaced other traditional cash crops, such as *dah*, yams and peanuts, as the first crop grown in the rotation, and is commonly followed by maize to scavenge any remaining fertilizer. Maize is another crop with a relatively recent introduction into the local field rotations. Although maize has been grown as a garden crop in West Africa for centuries (introduced as early as the 1600s; Miracle, 1966), when describing the cropping systems used by their parents, farmers in only one area indicated the use of maize in field rotations. Now, in nearly all locations of the Central and Southern Portfolio Areas, farmers include the cultivation of maize as a regular crop in their production systems.

The production of market-oriented fruit and vegetable crops is another area of activity whose importance has grown tremendously since the preceding generation. Although vegetable production had briefly been a focus of development assistance in the OHVN area in the 1960s, farmers in only three villages reported extensive dry-season gardening as a major source of income during the time of the preceding generation. Presently, tomatoes, onions, peppers and okra, among other vegetables, are major cash crops for farmers in nearly every part of the zone. For farmers in the Southern Portfolio areas the rise in importance of fruit crops is another significant change in their household production systems. Mangoes, citrus and bananas are common horticultural crops grown commercially in small, home garden plantations, while water-melon is a major, late season field crop. In some locations farmers are experimenting with less 'traditional' commercial species, e.g. pineapple and guava. In response to farmers' expansion of their market-gardening efforts, the extension programme issued a collection of technical *fiches* on vegetable production in 1993.

Home gardens, agroforestry and forest management

The recurring droughts that began in the 1960s forced many pastoral groups to move southward into areas of more intensive agriculture, where they have since settled. Their relocation has paralleled the general increase in both the adoption of animal traction and local investment in livestock in southern Mali (e.g. Foltz, 1991). Consequently, during the last 20 years farmers in all but the most arid areas of the north have responded by developing systems of 'living fences' to protect their crops from livestock. The most common species used as a living fence is *pourghère* (*Jatropha curcas*), although sisal (*Agrave sisalana*) is also common in the Southern Portfolio areas. *Pourghère*, which is inedible to livestock, is native to the American tropics and was most likely imported to West Africa by the Portuguese (Jones and Miller, n.d.) and spread during the colonial occupation.[19] Species used as living fences are now planted to form complete enclosures around home gardens or larger fields, as barriers along one or more sides of a field, and along the major pathways where livestock are taken into and out of the village.

Farmers report that the present generation was the first to design and manage the extensive home gardens and household plantations that are now characteristic of the Southern Portfolio Areas of the zone, especially in communities on the western side of the Niger River, which enjoy relatively easy access to the Bamako markets, and where commercial cotton production has historically been less important. The development of these home garden/plantations reflects the general increase in importance of market-oriented horticultural production, and has led to a concomitant increase in farmers' skills in plantation management. In the mid-1960s, a group of what farmers described as 'tourists' passed through the area and introduced them to tree grafting. Individuals have since greatly improved their grafting skills and now actively work to expand their holdings of genetic material and develop their own varieties; a common feature in most home gardens is a small nursery where farmers multiply seedlings both for personal use and to sell. Farmers with well-developed grafting skills are able to secure dry-season employment on the commercial plantations around Bamako. Despite the high level of interest in horticultural activities, the OHVN technical package has had little to offer small-scale plantation managers, although some field agents have recently begun to receive training in tree-grafting techniques.

In addition to plantation crops, agroforestry associations, as noted in Chapter 4, is another area of high activity in local experimentation. Although farmers throughout West Africa have a long history of protecting emergent seedlings of desired species around their villages, compounds and in their fields, documented historical examples of active tree planting are rare (e.g. Fairhead and Leach, 1996). In fact, farmers in many locations noted that the tree densities in their main production fields have been on the decline for several decades. In the northern areas this decline is generally due to the impact of droughts, while in the south farmers linked this decline to the increased use of the plough. As in the case of the abandonment of rocklines, farmers reported that few of the naturally emerging seedlings are being protected and nearly all of the living tree stumps are being removed from fields as they are taken out of fallow due to the difficulty in ploughing around them. And yet this situation is changing.

Farmers in both northern and southern localities are beginning to re-establish desired tree species in their main production fields. In a town in the Far North Portfolio area, one of the more affluent farmers was producing seedlings with assistance from the Eaux et Forêts (National Water and Forest Service), while others in this area and elsewhere in the zone are performing these activities completely unassisted (cf. Montagne, 1985/86). The OHVN extension service has begun encouraging farmers to plant neem, African mahogany (*Khaya senegalensis*) and a few other single-use species along field boundaries, but not as agroforestry associations within fields. The one exception is the experimental planting of *Acacia albida* in fields in one village in the Far North Portfolio area, which was encouraged by the DRSPR/OHV research programme. On their own, farmers across the zone have begun to multiply and plant a much wider range of multi-purpose species directly into their main production fields, as well as around their compounds and in established home gardens. Economic analyses show that the benefits derived from the secondary products of two of the most widespread agroforestry species, *karité* (*Vitellaria paradoxa*) and *néré* (*Parkia biblobosa*), outweigh the negative impact they have on crop production beneath their canopies (Boffa *et al.*, 1996; Kater *et al.*, 1992; Kessler, 1992).

Another relatively recent innovation is the creation of community wood-lots that have been increasingly promoted by governmental services and NGOs to solve local fuelwood shortages and help protect the remaining natural forest stands. In practice, however, these plantations have not secured the necessary community support for their success. The plantations observed during this study had either been installed entirely by an NGO, or communities had been persuaded to purchase and plant seedlings through governmental programmes. In both cases, seedling survival rates were low, ranging from 5 to 20 per cent due to heavy dry-season losses. Unclear tenure rights, benefit streams and the lack of attention paid to organizational issues related to plantation management have contributed to the poor results.

Beyond the continued protection of 'sacred groves,' traditional systems of forest-management have largely broken down. The historical loss of local control over forest resources through the colonial forest codes, maintained by subsequent governments, has played a primary role in the desiccation of many communities' forest reserves (e.g. Thomson, 1987; Thomson *et al.*, 1986; cf. Djibo *et al.*, 1991). In one community, however, people were attempting to use these same national laws to their advantage. In this instance, the community had applied for their forest lands to be granted *Forêt Classé* status, thereby protecting them from further destruction by outsiders and village residents whose extensive woodcutting has been supported by an NGO working in the area. Under the First Amendment to the USAID-financed Development of the Haute Vallée Project (USAID, 1993), selected communities were chosen for involvement in a pilot programme that will begin transferring forest-management responsibilities back to local communities in an effort to rebuild local management capabilities and evaluate the potential of local organizations assuming a greater role in active forest management (cf. Thomson and Coulibaly, 1995).

Animal husbandry and livestock associations

As noted, there has been a significant southward shift of many pastoral groups and a general increase in livestock investments in the southern half of the OHVN zone over the past generation. For a number of communities, the past few decades have brought noticeable increases in marketing opportunities for livestock, animal products (such as fresh milk and eggs) and fodder, which includes both the substantial sale of fodder to urban residents, as well as, in some instances, to passing pastoralists and rural producers. The increased importance of animal manure to most cropping systems has led to greater efforts being expended in transporting manure to the fields by carts, or constructing traditional temporary corrals in fields so that manure can be deposited *in situ*. Toulmin (1991; 1992) describes the changes in exchange relationships among agriculturists and herders, and the growing importance of infrastructure investment (well construction) for ensuring a continued supply of manure. The benefit–cost ratio of investments in improved livestock corrals promoted by the research and extension programmes, however, appears questionable, while the substantial outlay of capital required to adopt the technology is beyond the means of even the wealthiest farmers (McCorkle *et al.*, 1993).

Other changes

Through the strong support of the research and extension services, private-sector interests and some NGOs, the past decade has seen a significant increase in farmers' use of purchased, inorganic inputs. The current use of fertilizers, pesticides and herbicides (in descending order) are most commonly associated with cotton and, to a lesser extent, maize and sorghum production. In general, fertilizer use has played a minor, but expanding, role in the production of cereal crops, and is most common in the better watered southern areas.[20] Use of insecticides and herbicides is much lower. The results of a seven-year study on insecticide use in Mali shows that the use of insecticides, even with the ultra-low volume sprayers, is not cost effective for millet farmers in the Sahelian zone (Jago *et al.*, 1993). Although external inputs are used mainly by men (due largely to their control over the formal lines of credit), women farmers in one area of the South-east have come to rely upon herbicides, obtained from their husbands, for their personal plots. Farmers' use of pesticides and herbicides include both standard application procedures and the mixing of biocide 'cocktails,' where a number of different chemical agents are combined and applied in a single dosage.

To protect their stored grain, farmers throughout the zone have come to use a variety of methods: purchased chemicals, ashes, the dried leaves of the neem tree and several 'bitter' herbs (e.g., *Benefin* (*Hyptis specigera*), a member of the mint family, and the dried fruit of the *Samakara* (*Swartzia madagascariensis*). While the use of specific insecticides for storage have been encouraged by the extension service, farmers in some locations reported using inappropriate commercial insecticides, and even herbicides, in their granaries for protection against pest loss. These spontaneous innovations have potentially tragic consequences. Although farmers reportedly 'cleaned off' the poisons before consuming the grain, the observed safety standards in handling these toxic substances, by both farmers and suppliers, were rarely adequate to ensure user safety. It is not surprising that a growing number of cases of pesticide poisoning have been reported (e.g. Kremer and Sidibé, 1991). Such incidents can be expected to increase with the rise in prevalence and use of agrochemicals, especially as some of the recommended treatments being promoted by the extension service include highly toxic substances (e.g. *Sijolan Rouge*, which contains known carcinogens—chlordane and hetachlor) (OHVN, 1993b; Anon., 1993).[21]

Neem leaves are another product that appears to have been introduced fairly recently in many villages. The neem tree, native to India, was introduced to Ghana as a shade tree by an officer in the colonial administration in 1919 (NRC, 1992), and is now an established part of the Sahelian landscape. The use of its leaves in protecting stored grain, while common across India and areas of West Africa (e.g. McCorkle *et al.*, 1988; Radcliffe *et al.*, 1995), is uneven in the OHVN zone. Farmers in some villages claimed that the use of neem leaves is a traditional practice, while others have only recently adopted it, and still others have never heard of its use.

Evolution in agricultural problems

The hierarchy of major agricultural problems, as perceived by farmers today, are quite similar to those reportedly faced by the previous generation. Various pests, weeds and rainfall top the list, though soil fertility, erosion and *striga* infestation are all more prevalent than in years past (see Table 6.1). With the exception of birds, wildlife seems to be of less concern in most areas, corresponding to the general decline in the region's fauna populations (Warshall, 1989). Problems with bird predation, on the other hand, are reportedly as severe, if not worse, than in the past. Farmers suggested that the lower rainfall supports a less abundant store of wild food, forcing birds increasingly to turn to farmers' fields for their meals. In villages where faster maturing crop varieties are just being introduced, the less extensive plantings serve to concentrate the bird damage that would normally be dispersed over a greater area; this has been a particularly serious problem with

Table 6.1 ***Farmers' principal agricultural production problems: past and present***

Past		Present	
Men	Women	Men	Women
1. Locusts	1. Birds	1. Birds	1. Birds
2. Birds	2. Insects	2. Insects/Storage/ Erosion	2. Insects/Storage/ Fertility
3. Monkeys/Insects/Weeds	3. Monkeys	3. Rainfall/Striga/Fertility	3. Striga/Locusts/Erosion
4. Pigs (wild)/Striga Drought/excess	4. Locusts	4. Locusts	4. Rainfall/Weeds/ Equipment/Animals (domestic)
5. Manual labour/Mice Disease (crop)	5. Mice	5. Weeds/Equipment/ Livestock	5. Land/Labour/Time
	6. Pigs (wild)/Striga Disease (crop) Equipment/Drought	6. Animals (domestic)	

millet. It has also been suggested that in some areas the early maturation of grain may also coincide with the migration patterns of some of the major pest species, thus making them targets of an even larger transitory population.

Summary

The examples of major agricultural change presented in this chapter help to illustrate the relative nature and extent of impact of the various sources of innovation on the evolution of household production systems across the OHVN zone. While the attribution of specific changes to specific 'sources' is in many cases rather tenuous, and leaves entirely untouched the issue of the relative value of different changes, the historical approach taken in this research is instructive in helping to identify some of the broader patterns of change taking place in the OHVN zone. The formal system of research and extension, for example, has had its greatest success in supporting the introduction of technologies, such as animal traction, use of inorganic inputs and the improvement of certain crop varieties, that are either based upon materials or derived through procedures which lie outside of farmers' ranges of experiences and resources or, in the case of cotton, are part of an integrated supply and retrieval system that greatly reduces farmers' transaction costs and uncertainty. The formal system has also had some success in refining and reintroducing certain indigenously developed management practices, such as the use of rock lines, ridges and the application of animal manure and compost.

Overall, however, it is quite apparent that farmers' own informal efforts have served as the major force in generating the continual stream of improvements and adjustments in the existing production systems that have allowed households to keep pace with their changing physical, demographic and economic environments over the past several decades. Farmers' efforts include the generation of entirely new management practices and productive enterprises, as well as the 'salvaging' and modifying of useful bits of information or adapting whole technologies developed by the formal system to fit their specific needs and conditions. In fact, the 'second-hand' involvement of farmers in refining many of the technologies developed by the formal system has greatly expanded the level of contributions made by the formal system to the evolution of local farming systems. The next chapter will focus on identifying some of the social and behavioral patterns that underlie farmers' generation and exchange of knowledge and innovations.

7 A conceptual framework for understanding the behaviourial dimensions of local agrarian change

This research was carried out using the Grounded Theory approach (Glaser and Strauss, 1967; Strauss and Corbin, 1990) (see Chapter 1) to help both broaden and sharpen our understanding of the behaviourial dynamics and processes operating within the 'black box' of local agricultural change. While the empirical observations and findings presented thus far have focused upon farmers in the OHVN, what is needed at this point is to re-examine the case-study for lessons and patterns of individual and social behaviour that can help to inform development activities within a much wider context. In taking this next step in the Grounded Theory process, this chapter reinterprets the research findings from the OHVN case-study (Chapters 3–6) using a synthesis of theoretical concepts and postulates drawn from the extensive knowledge base on human behaviour and social interactions, which, to date, has been little used in assisting agricultural development activities.

In addition to constructing the beginnings of a framework to guide future developmental action, it is also hoped that the use of this material will help (in part) to demystify the behaviourial science perspective and provide a point of entry for the further utilization of this rich body of insights in better understanding and building upon the essential contributions of the social and individual aspects of agricultural change. The conceptual perspectives used in this chapter were drawn from the literature of social and personal construct theory, cognitive and social psychology, cultural anthropology, and various learning and communication theories, as well as the writings of a number of contemporary development scholars. The chapter is divided into three parts, beginning with a general overview of the theoretical relationships between culture, society and the individual, and the introduction of a basic template for understanding the social and individual dimensions of knowledge differentiation. In the second part, important characteristics of local knowledge, and the different states, forms and levels in which knowledge is found, are introduced. Finally, a basic model for understanding innovation-based (agri)cultural change is introduced—one which emphasizes the dynamic interplay between cultural traditions and the processes of individual creativity, learning and communication that enables cultures to evolve and sustain themselves.

Culture, society and the individual

In exploring technological change among 'traditional societies', Foster (1973:10) structures his investigation around the interconnection of 'three systems': the *social* (distinct groups of people and their structured relations), the *cultural* (the shared value-system and normative behaviour of groups) and the *psychological* (individual cognition and motivation). Although it was not the intent of this research to describe and study the functioning of these systems among communities in the OHVN, the collected data allow, and in fact require, some general remarks to be made concerning the social, cultural and psychological influences that help to frame the 'central issue' facing behaviourialists, 'the search for autonomy in the midst of [social] constraint and the countervailing search for control in the face of [individual] licence' (Bennett, 1976:852).

Viewed in general terms, individuals can be seen as active participants in a rich drama of creating, maintaining and adapting their social relationships, beliefs, values and behaviours—those things that constitute a group's socio-cultural heritage and provide both a sense of identity and security in 'belonging'. In helping to preserve a group's identity, socio-cultural institutions, as the 'complex of norms and behaviours' of the group (Uphoff, 1986:6), provide individuals with two important services: the source of positive and negative reinforcement of acceptable and unacceptable behaviour; and an influence on 'the process of individual learning'—a process which 'consists of adopting highly specialized attitudes toward reality... so deeply implanted that it never occurs to the individual that there might be any others' (Towner, 1980:171; cf. Schutz and Luckmann, 1973). These 'attitudes', embedded in the codified symbols and language of the group, provide some of the basic templates and guides through which individuals filter and impart meaning to the day-to-day sensory information which they encounter. Although individuals do not hold identical visions of what reality is, through the interplay between the individual and group 'there is an ongoing correspondence between *my* meanings and *their* meanings', such that there develops a perception that '*we* share a common sense about... reality' (Berger and Luckmann, 1966:23).

As observed in the case-study, there are vast differences between the world views and perceptions of reality held by rural farmers and members of the formal research and extension system. These differences range from the mundane (e.g. the basic definition of units and nomenclature of classification systems), to the spiritual, where, for example, farmers' world views (or cosmovision) include the existence and involvement of spiritual forces in land degradation and field performance, and the interpretation of dreams and phases of the moon as guides for when to undertake certain agricultural practices. Such explanations and sources of guidance are vastly different from those recognized as valid by researchers and extensionists, and are indicative of the underlying differences in the socialized values, beliefs and norms of behaviour, life experiences and patterns of reinforcements that are active within these two groups.

The cultural-centric perspective that develops as part of the socialization process reinforces the need for a multicultural framework for identifying and addressing issues of human behaviour. A diagrammatic tool, the 'Johari window', borrowed from the study of personal awareness in psychology (Luft, 1970), is modified in Figure 7.1 to help depict how individuals from different social systems and perceptions of 'reality' possess different knowledge, and, in fact, often have a limited awareness of each other's perspective and understanding of how the world operates. In contrasting the technical knowledge held by researchers and farmers, this figure illustrates that while certain things are known to both groups, other elements are known only selectively (or are known in very different ways), and some are known to neither. The obvious implication from a knowledge-systems perspective is the long-argued position that both researchers and farmers have unique contributions to make, and that neither 'system' can fully replace the other. In fact, one could argue that each 'system' needs the other if the full store of human potential is to be utilized. As will be stressed later, this same perspective can also be used in arguing for the inclusion of different disciplinary perspectives from within the formal research system, as well as the inclusion of different 'groups' and individuals at the local level.

		Researchers	
		Known	Unknown
Farmers	Known	Commonly held knowledge	Unique local knowledge
	Unknown	Unique scientific knowledge	Unknown

Figure 7.1 *The intersection of informal and formal knowledge*

Social and personal constructions of reality

As the group and individual interviews revealed, farmers possess elements of agricultural knowledge that are both broadly shared and much more personal in nature. In moving towards a better understanding of the intra-group similarities and differences in knowledge formation, the introduction of two contrasting theories provides a useful point of departure. In the social constructionist view, reality is a normatively defined, socially replicated and reinforced perspective, based upon the postulate that the joint definition and belief of something as real leads to it becoming 'reality as subjectively experienced by the members of that society' (Thomas, 1923). This view is not concerned with providing commentary on the existence or nature of an ultimate reality. Rather, the focus is upon the way in which reality is jointly defined and experienced by social groups at a particular point in time (e.g. Berger and Luckmann, 1966; Childe, 1956; Holzner and Marx, 1979; Knorr-Cetina, 1981, amongst others). While the process of (re)constructing a joint perspective can be fraught with contention, at the centre of the constructionist view is an evolving core of knowledge and belief that is rarely, if ever, questioned and which it is assumed that everyone shares. An example of this type of 'unquestioned' knowledge from the case-study appeared in Chapter 4 (Box 4.1), where in reply to interview questions regarding the different natural signs that were used in determining when the rainy season had arrived, the farmer responded: '...what we know about birds, well, that's something that everyone knows. So if I skipped over that in the first place, it's because it's common knowledge' (McConnell, 1993:Appendix A).

The psychological counterpart to the social constructionist perspective is found in the 'personal construct' psychology proposed by Kelly (1955). Consistent with the focus on individual cognition inherent to psychology, the central premiss of Personal Construct Theory is that individuals function as their own 'personal scientist', driven by the desire to understand and predict—'to devise conceptual templates [personal constructs] that permit them to interpret, anticipate, and appropriately respond to the events with which they are confronted' (Neimeyer, 1985:2). In this view, human understanding is fuelled more by individual inquisitiveness, creativity and choice, than by social conformity. Examples of this type of knowledge are found in the rationale used by individual farmers in making specific management decisions, such as the selection of a particular intercropping pattern, crop rotation or variety over another. These elements of knowledge and belief are, by definition, not widely shared, but are used by individuals as guides in making many of their daily management decisions.

In attempting to integrate both personal and social knowledge constructs, it is helpful to first look at the underlying architecture of knowledge within rural communities. At the broadest level, every community or society can be said to constitute a 'social pool' of knowledge, composed of the total stock of knowledge held by that society's members (Berger and Luckmann, 1966; Schutz and Luckmann, 1973). By definition, this 'social pool' far exceeds the personal stock of knowledge held by any individual, or collection of individuals, within a community. The 'social pool' in turn contains a 'common stock', or cultural core, of socially constructed and shared knowledge, held more or less equally by all members of a society, and often defined as literally common-sense knowledge (Berger and Luckmann, 1966; Holzner and Marx, 1979) (see Figure 7.2). This common-sense, or common stock, knowledge includes the social rules, behaviourial expectations, shared values, beliefs and specific elements of practical knowledge that provide the medium of social intercourse evident in the normal rhythms of everyday life. It is, in effect, 'believed to represent reality itself' (Holzner and Marx, 1979:259). In addition to this common stock of knowledge is the knowledge shared on a more limited basis among members of sub-groups within a community, as well as the unique knowledge holdings of individuals. These added dimensions of knowledge differentiation can be explained by the various 'social frames of reference' with which individuals are associated, and the personal, experientially based, knowledge 'biography' that every individual amasses during the course of her or his life-time.

Social frames of reference and individual biographies

The patterns of group and individual responses that emerged during the interviews indicate the existence of some important qualitative and quantitative differences in the knowledge held by individuals from similar socio-cultural backgrounds. While these differences generally fall short of defining separate knowledge networks, or 'epistemic communities' (Long and Villareal, 1994), the differences between, and internal linkages within, certain religious, ethnic and caste groups, could support such an argument. Overall, as depicted in Figure 7.2, the social differentiation of local knowledge is rooted in the socially defined roles and status, and their associated norms of behaviour, through which individuals identify themselves and one another, and manage their relations within the community. This type of knowledge differentiation may be best understood through the identification of the different 'social frames' (Gurvitch, 1971), or joint 'frames of reference', with which individuals are associated (Holzner and Marx, 1979). As described in Chapter 4, these social frames range from those associated with the relatively 'closed' (ascribed) aspects of lineage, gender, caste and ethnic divisions, to the more 'open' or achieved levels of social status associated with such factors as age, wealth and religious affiliation. By influencing an individual's access to resources, performance of specific tasks and the general pressure to follow certain patterns of behaviour, the combination of social frames with which an individual is associated serve to delimit (enabling as well as constraining) her or his personal horizon of real-life experiences and, consequently, much of her or his acquisition of experientially derived knowledge (e.g. Schutz and Luckmann, 1973; Rosenthal and Zimmerman, 1978). Viewed in such a way, a social frame perspective provides a useful tool for understanding, at a general level, some of the dynamics involved in knowledge formation, as well as providing the basis for explaining the patterns of similarity and differences found

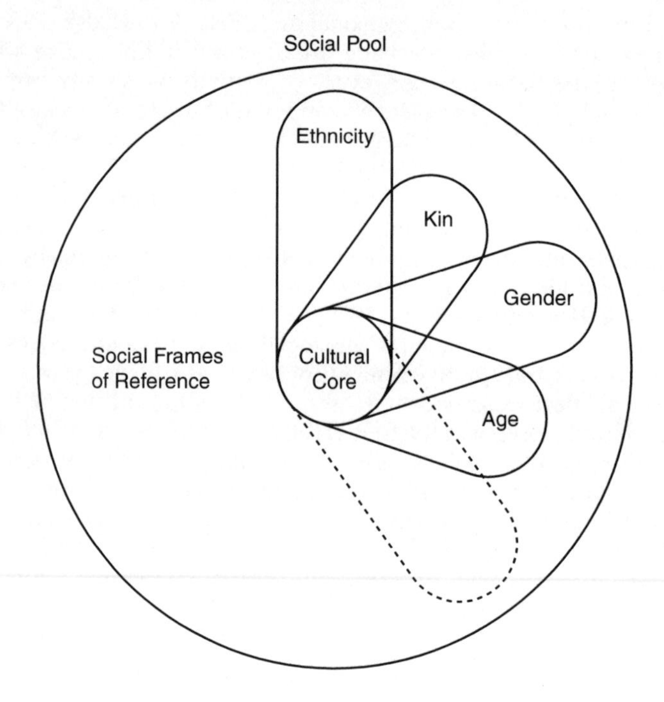

Figure 7.2 *Features of local knowledge systems*

within the knowledge holdings and interpretive meanings of individuals within the same communities.

Viewed from a methodological perspective, the social influences on an individual's acquisition of knowledge about different things, different aspects of the same things, or in interpreting and organizing the same information in different ways (Norem *et al.*, 1989), highlights the importance of using iterative rounds of inquiry and engaging a wide range of individuals in data collection and development activities. While the caution raised by Fairhead and Leach (1994) over the danger of assuming, a priori, that differences exist in the knowledge held by individuals from different gender, age, ethnic and class groups needs to be taken seriously, the observations from the OHVN clearly show that a number of significant differences do exist and that, once identified, such differences must also be taken seriously and explored.

In addition to those social frames that were found to influence the differential acquisition of local knowledge (e.g. kinship, gender, ethnicity, caste, age and wealth, elaborated in Chapter 4), the field data also highlighted the existence of important individual differences among members of similar social frames. Such differences are based upon variances in personal experiences (e.g. education, travel), creativity, skill and motivation; those unspoken and unseen differences that allow one person to 'make frail plants flourish where others only have to raise a watering can for them to die' (Richards, 1993:62). If a 'social frames' perspective is to help explain this second axis of knowledge differentiation, it first must reject the type of 'determinism of existing general theories of social change' (Long, 1984:171) that anthropologists have long found to have 'contributed to a cumulative distortion in our image of the practice of traditional agriculture'

(Johnson, 1972:151; cf. Fairhead, 1993). What is needed is 'to combine structural analysis [i.e., a social frames perspective]... with an actor-oriented approach' (Long, 1984:175; cf. Giddens, 1979; Long and van der Ploeg, 1994). In other words, a perspective that views individuals as 'situated-actors' (Bebbington, 1994b; cf. Giddens, 1979), possessing knowledge that is not only based upon their socially influenced range of experiences, opportunities and behaviours, but also as a product of their individual inquisitiveness, cognition and decisions.

In further exploring this type of 'actor-generated' knowledge, the concept of 'individual biographies', introduced by Berger and Luckmann (1966), provides a useful temporal dimension for understanding the idiosyncratic differences in personal constructs, knowledge and behaviour encountered during the field work.[1] From this perspective, each individual's knowledge represents a 'biography', or living record of one's life history (see Box, 1988), resulting from their unique combination of personal attributes and life experiences. Whereas the socialization process tends to produce a shared outlook and understanding of the world, each individual's personal knowledge develops as an outgrowth of their life-time of observing, doing and learning—elements of which may in turn influence the 'group.' Such a view is consistent with the broader theoretical notion of 'reciprocal determinism', as outlined in Bandura's *Social Learning Theory* (1977), where 'human functioning neither casts people into the role of powerless objects controlled by environment [of which social forces are a part] nor free agents who can become whatever they choose. People and their environments are reciprocal determinants of each other' (Bandura, 1977: vii).

The nature and organization of knowledge

Because knowledge is one of the essential elements of culture—the complex of shared beliefs, values and behaviours that defines a way of life and renders each group distinct from its neighbour—the temptation has often been to view all local knowledge with a high degree of uniformity and orderliness. However, as the research findings and preceding discussion illustrate, the portrayal of local knowledge as a singular, highly organized system is misleading in that it implies a greater degree of homogeneity in individual 'knowing' and fluidity of information exchange than actually exists (e.g. Swift, 1979; Scoones and Thompson, 1994). While the social and individual differentiation of knowledge could lead one to question the very existence of local knowledge 'systems' altogether, the implication that a systems view should be reserved solely for Western or Eastern 'science' is flawed in its exaggeration of the differences between local and formal knowledge (cf. Agrawal, 1995). As the examination of formal research institutions reveals, the differentiation of individual knowledge appears to be part of the nature of knowledge, even within the formal scientific community (e.g. Knorr-Cetina, 1981; cf. Bijker *et al.*, 1987). Each individual's knowledge, for example, is differentiated by disciplinary views, increasingly divergent schools of thought, the organizational culture of each research facility, personal observations of each researcher and successive revolutions in scientific thinking involving what, and how, knowledge is agreed upon. Some of these differences can be seen among the different farming systems research units active in Mali, where differences in staff profiles, proximity to other research departments and the different sources of technical assistance that they have received (Dutch versus US) have led to the utilization of very different perspectives, approaches and specific methodologies in their work.

Within the realm of agriculture, one area where the 'formal' and 'informal' systems do show a great deal of divergence is in the organization and use of knowledge—in particular, the extent to which technical knowledge is separated from the rest of the knowledge holder's beliefs and activities. Members of the formal system, for example, are employed specifically to operate an explicit (normative) system of rules and procedures in generating new information. These individuals, however, are generally not the intended end-users of the information that they produce. Their immediate livelihoods and well-being, and that of their families, are not dependent upon the utility of the knowledge that they generate in terms of practical application. In other words, there is a certain separation, and 'cushion', between the knowledge system of their profession and the other areas of their lives. Farmers, on the other hand, generate and utilize different sets of 'practical knowledge' (Sternberg and Wagner, 1986) that are central to their immediate survival. The elements of technical knowledge that farmers utilize are inextricably linked to a range of social processes, such that the technical 'knowledge system does not appear as an empirically separate aspect of the social structure' (Holzner and Marx, 1979:175), but is an essential part of the '...continuing [social and technical] adaptation and survival of the group in the ecological setting in which its members find themselves' (Berry and Irvine, 1986:271; cf. de Schlippe, 1956). Even though farmers and scientists may use similar approaches in conducting their research (e.g. Potts *et al.*, 1992), no explicit rules govern the generation of knowledge by farmers. Information is validated through its utility in practice. Seen from a knowledge user's perspective, the perceived relevance of knowledge to the successful performance of specific activities is of greater importance than is the overall comprehensiveness or fit of that knowledge as part of a universal explanatory system.

Forms, states and levels of knowledge

Apart from its relationship with specific activities, knowledge within human consciousness can be said to reside in a number of forms and states, and to exist at several levels of inclusiveness. Two basic forms that knowledge can take are explicit knowledge (that which is consciously held, or integrated into an individual's larger set of understanding), and implicit knowledge (that which is subconscious, or as yet unintegrated into one's understanding) (Cleeremans, 1993; cf. Gladwin and Murtaugh, 1980). Knowledge also resides in one of two different states—tacit (unspoken), and that which is openly communicated—and exists at a number of levels, ranging from knowledge 'of', or existence knowledge, knowledge 'how', or performance knowledge, and knowledge 'what', or process knowledge, which includes an understanding of the dynamics or mechanisms by which something works (adapted from Scoones and Thompson, 1992). Based upon findings represented in Chapters 4–6, Figure 7.3 illustrates these different dimensions of personal knowledge in terms of the forms, states and levels that knowledge can take. For example, irrespective of the depth of our understanding, the simple fact that we often 'know more than we can tell' (Polanyi, 1966:4) illustrates the existence of both implicit/tacit knowledge, and helps to explain the difficulty farmers often have in articulating things that they know, yet which have neither been fully integrated into their conscious understanding of how the world operates, nor been previously communicated. At other times we also 'tell more than we realize' (implicit/communicated), a situation diametrically opposed to those in which we attempt to teach others specific things (explicit/

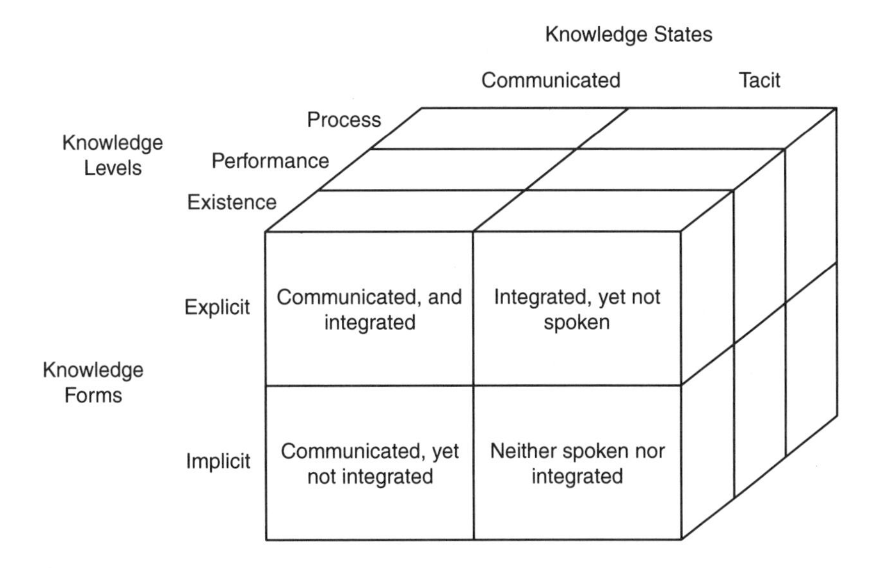

Figure 7.3 *The states, forms and levels of personal knowledge*

communicated), or when we possess bits of explicitly held knowledge that we could pass on to others but, for whatever reason, choose not to (explicit/tacit) (Berger and Luckmann, 1966). These different dimensions of knowledge illustrate the necessity of employing a range of methodologies and practices in working with farmers to help them identify and communicate their knowledge.[2]

The movement of knowledge between the different forms, states and levels of inclusiveness, when combined with individual differences in access to information and experiences, as well as intellectual capacities, interests and motivations, underscores the point that individual knowledge biographies are 'fragmentary, partial and provisional in nature... never fully unified or integrated in terms of an underlying logic or system of classification' (Arce and Long, 1992:211), and as a consequence are '...always in the making' (Scoones and Thompson, 1992:4). The important lesson here is that, individually and jointly, our breadth, depth and degree of integration of understanding are continually evolving.

Agricultural change and cultural evolution

In considering the processes of (agri)cultural change, it is important to first recognize that all knowledge, even that which is socially (re)constructed and culturally shared, 'is first acquired by individuals' (Towner, 1980:171). Although change within a cultural system is 'initiated by individuals' (Barnett, 1953:39), it is also important to note that no two stimuli are exactly alike, and that no two individuals respond to the same situation in exactly the same way (ibid, 1953:19). Thus innovation can be conceived as an individual psychological process, with social consequences related to the extent of an innovation's acceptance, spread or modification within the larger social group (ibid, 1953:1). The movement of ideas, materials and innovations within a culture, from the individual to the larger social group, and the exchanges made with external sources (depicted in Figure 7.4), constitute the essential processes involved in the technical and social evolution of (agri)cultural systems, the tempo of which changes in response to varying internal and external forces.[3]

The accumulation of knowledge within individual biographies takes place through a combination of processes involving the acquisition of pre-existing knowledge[4] through various channels of socialization (communication, observation and action), as well as the generation of new information through personal experimentation and discovery. When viewed over time or, as in the OHVN case study, by collecting data covering various time periods, the evolution and maintenance of culturally shared knowledge can be seen as a process fuelled by individual creativity and inter-personal communication (inter- and intra-generational, as well as between members of different cultural groups). These processes of creating, reproducing and exchanging information and materials can be described as taking place among three main sources—the individual, cultural and external. Shown in Figure 7.4, the description of and interactions between these three sources represent a simplified model of the major interfaces among the different reservoirs of knowledge and sources of information important to knowledge-based change in agricultural production systems.[5] At the individual level, the degree of overlap between these three spheres of knowledge varies from person to person, depending upon the breadth and depth of one's biography of experiences, association with various social frames of reference and personal contacts (i.e. embedding the features depicted in Figure 7.2 within the processes of change modelled in Figure 7.4). Each individual draws upon, helps to preserve, revises and contributes to the collective knowledge pool of her or his culture, as well as the knowledge pools of groups outside their immediate socio-cultural context. As seen in some of the examples presented in Chapter 6, (agri)cultural traditions evolve as innovations, developed by individuals within the society or brought in from the outside, become widely diffused and integrated into the current practices and are passed on to the next generation as part of their cultural heritage. Broader culture to culture exchanges occur through such avenues as mass media campaigns, educational programmes and the re-shaping of the economic and political playing fields via policies that affect the opportunity sets of entire segments of a society *en masse*.

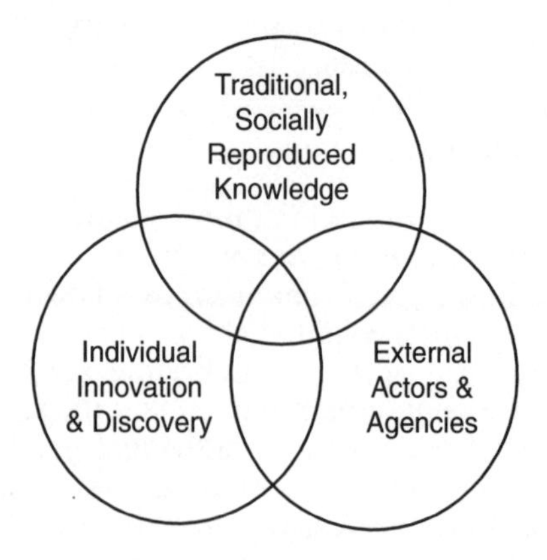

Figure 7.4 *Sources of information and frontiers of interaction in the evolution of local knowledge systems*

The evolution of technical knowledge is characterized by the inherent state of tension, or 'conflict between the freedom to vary, which makes advance possible, and the value of retaining the cultural accumulation' (Campbell, 1965:35; cf. Bennett, 1976; Buckles and Perales, 1997; Foster, 1973:77; Vanberg, 1992). This continual interplay, or creative friction between cultural traditions and individual variation, highlights the notion that traditions embody a 'living past', while at the same time a certain degree of 'change is internal to tradition' (Zaretzke, 1982:91); an important observation, since all cultures possess 'traditions of creativity' (Silverman, 1994), as the following example from Ethiopia illustrates:

> ...inquiries about the origins of various [milk] container designs were often answered with references to *aada*—the idea of cultural heritage, traditions that are passed from generation to generation... At first glance, Borana milk containers might appear very similar if not identical. But not all woven milk containers are alike—in fact, because they are each handmade they are all different and there is a good deal of room for innovation. A woman may experiment; she may use new materials, alter the proportions or profile of a container, or introduce new surface treatments. If accepted and copied by others, such a creative act by an individual may be integrated into the tradition and thus become *aada Borana*—part of the Borana heritage (Silverman, 1994).

As noted in the preceding discussion, the central forces in the evolution and maintenance of (agri)cultural knowledge systems are learning and communication. Learning involves both the reproduction of cultural traditions, and the creation of new knowledge through innovation and discovery, while the transfer (communication) of existing and new information and skills from one individual to another, and one generation to the next, constitutes the vehicle through which cultures endure, and the process whereby 'today's innovations become tomorrow's traditions' (Wilson, 1984).

Learning, creativity and experience

The number of theories on learning in the behaviourial science literature are legion. As seen in the case-study, and depicted in Figure 7.4, farmers learn from a number of sources (the preceding generation, each other, outsiders, and by doing). For the present, the discussion will focus upon those types of learning associated with individual innovation and discovery encountered in the field work. At some point, 'those who concern themselves with innovation begin with, or come to accept, one of two antithetical propositions relative to human inventiveness. It does not matter whether their point of view is founded upon general impressions or careful study; in either case they find themselves impelled to believe that human beings are fundamentally creative, or that they are not' (Barnett, 1953:17). The observations drawn from the OHVN case-study unequivocally support the proposition that individuals *are* inherently creative beings. The very nature of the world around them requires that farmers be 'creative' in order to adjust to the evolving nature of their physical and biological environments, as well as the social, political and economic systems.

As seen in the case-study (Chapters 3, 4 and 6), innovation is a creative response often arising from the necessity of, desire to, or curiosity about, change. The moment of an innovation's conception has been likened to a 'lightning bolt' of insight (Knorr-Cetina, 1981), where previously ordered concepts and empirical evidence are recombined to form new results, or are extended in taking the next 'logical' step (often guided by intuition), or are transferred from one context and applied within another. Kuhn (1970), for one, notes that many major breakthroughs in formal science are made by people who are young and/or new to their fields, whose thinking is not yet bound by the standard perceptions and who may bring knowledge from other areas to bear on the problems at hand. A similar set of processes can be seen to operate among farmers who, for example, have been educated or raised in urban centres and have since returned to their natal village and begun to experiment with entirely new enterprises and management systems. At other times, innovations arise from the observation and recognition of the utility of exceptions to the norm. Richards' (1985; 1986) description of the practice among Sierra Leone rice farmers of 'roguing' unique rice cultivars from their fields for separate experimentation, in order to identify their potential as new varieties, is one well-known example. Other innovations emerge from accidents or serendipitous discovery (Barnett, 1953; Knorr-Cetina, 1981:61; Rhoades and Bebbington, 1995; Stolzenback, 1997).

Learning theories, and virtually all writings on cultural change and human adaptation, emphasize the importance of experience and the learning environment. This experiential person–environment interface not only provides the opportunity for individuals to make observations and discoveries used in building and reforming their (explicit) explanatory theories and concepts about how the world operates (Kruglanski, 1989; Towner, 1980), but the continuous interaction between the individual and her or his physical and social environment also provides the opportunity for acquiring an implicit, or pre-attentive, understanding of how the world works (Gladwin and Murtaugh, 1980; Holzner and Marx, 1979). Farmers' lifelong contact with their local environments provides ample opportunity to develop extensive stores of both explicit and implicit types of knowledge.

Communication and the diffusion of information

In addition to the role of creativity and innovation, the case-study (Chapters 4 and 6) also provides a number of examples illustrating the importance of communication (among kin and religious groups, friendships and labour relations) in the 'vertical' and 'horizontal' learning that is constantly taking place within and between communities.[6] One attribute of the social frames of reference perspective described previously (and as outlined in Chapter 4) is that they not only shape individuals' opportunities for observation and personal experience, but they also strongly influence interpersonal lines of communication, e.g. who talks to whom, which is further tempered by personal affinities. Individuals acquire knowledge from others through overt instruction and demonstration (e.g. age-mates and family members), as well as through the tacit transfer of specific skills and competencies communicated through periods of guided experiential learning and careful observation (Holzner and Marx, 1979). Most social learning theories (e.g. Bandura, 1977; Rosenthal and Zimmerman, 1978) in fact posit that much of our learned behaviour is acquired through the explicit and implicit modelling of others.

There can be no question as to the central importance of social relations and communication pathways to the maintenance and evolution of (agri)cultural knowledge systems. As a process, however, knowledge transfer (even explicit attempts) is always imperfect, conveying less or interpreted differently than the communicator or receiver desires. Thoughts are often poorly expressed, contain incomplete information, or lack the contextual detail necessary for their full comprehension. This is especially true of knowledge that is just emerging from tacit storage, or implicit cognition which has not previously been articulated and opened to social commentary (Cleeremans, 1993). Conversely, information is also imperfectly received in terms of both content and meaning, dependent upon the attentiveness, interpretive skills, and personal biographies involved, as well as the embedded authority and credibility of the source (Havelock, 1973; Walter, 1993; Wilson, 1983). In the 'demi-arc' communication model, proposed by Haray and Havelock (1971), information transfer is defined as a 'meshing' of the sending and receiving halves. The sender determines what and how selected information is to be encoded, while the receiver must decode and interpret this information, 'forming connotations with schemata and memorized experiences and relating it to knowledge he or she already has' (Mundy and Compton, 1995:112; cf. Dissanayake, 1986).

Although diffusion theory has fallen out of favour in the 1990s, several of its core ideas on the role played by interpersonal relationships in moving information and materials across the different social interfaces remain valid. The major features in the social differentiation of knowledge, detailed in Chapter 4 and conceptualized here by the notion of social frames of reference, are analogous to what Rogers (1983) termed 'homophilous' conditions, which denote the degree of similarity between individuals' social experiences and the likelihood that these similarities will serve to facilitate the exchange of information between individuals. The observed importance in the transfer of genetic material and new technologies at local and regional markets, within certain forms of group labour (e.g. age-set *tons*), as well as at other activities and locations that serve to bring together individuals of different kin and ethnic groups, is captured by Rogers (1983; cf. Granovetter, 1973) in the notion of the 'strength of "weak" links'—where such types of relationships are weak in terms of 'homophilous' similarity, yet strong in terms of their capacity to introduce new materials and information. The maintenance of long-distance relationships, such as in the case of the OHVN where women and others visit their villages of origin, constitutes a slightly different version of the 'strong "weak" link' perspective (homophilously strong, yet weak in terms of frequency of contact), yet which is equally important in the introduction of new seed varieties. In general, the observations from the OHVN case-study, as well as investigations from other regions, show that, for rural villagers, project managers and post-industrial applied scientists alike, interpersonal contacts and informal lines of communication are among the most important sources of information regarding practical details (e.g. FAO, 1995; Holzner and Marx, 1979; McCorkle *et al.*, 1988; Nazhat and Coughenour, 1987; Nelson and Hall, 1994; Okali *et al.*, 1994a; Richards, 1994; Sumberg and Okali, 1997; amongst many others).

In instances where actors from different social groups or cultural contexts attempt to exchange their knowledge, an additional set of parameters comes into play at the 'exchange interface' (see Long, 1989). Chambers (1983), for one, has long noted the many biases affecting relationships between 'insiders' and 'outsiders' in the standard approaches to rural development. Other factors, such as linguistic nuances (see Chapter 4) and radically different world views, can confound even the most careful attempts at interacting with individuals from other cultures. With reference to the Johari window depicted in Figure 7.1, such factors highlight the need for a greater degree of sensitivity to local people's perspectives and, at least initially, the need to adopt a slower pace to allow for inter-cultural learning than is common in most development approaches.

Summary

The theoretical interpretation of the research findings presented in this chapter is intended to serve as a lens to help improve the clarity of our understanding of the many individual and social processes involved in local agrarian change. It is hoped that the insights gained will provide a basis for making a number of improvements in the orientation, objectives and methods used in agricultural development activities. Taken as a whole, the synthesis of concepts drawn from the behaviourial science literature forms the beginnings of a framework for understanding the processes of change at the local level. This emerging framework highlights a number of important features in the structure and dynamics of local knowledge systems: the existence of both normative and individual elements of understanding; the influence of social frames of reference and accumulation of individual knowledge biographies; different states, forms and levels of knowledge; and the essential contributions of creativity and discovery and informal lines of communication to individual learning and, ultimately, cultural evolution. Methodologically, the recognition of differences between the world-views of members of the formal system and farmers, as well as differences both among individuals and members of sub-groups within local populations, is a crucial starting point for formal research and extension organizations interested in engaging local knowledge and creativity. Not only do these features of the local knowledge system speak to the need for including multiple individuals from different social frames (or at least a conscious recognition of which social frames are represented or excluded), but the different forms, states and levels of knowledge will require the use of different approaches for accessing information from these different stores, as well as efforts to build a shared basis of understanding.

At a more fundamental level, this review calls for a deeper appreciation of the central role played by individual creativity and communication in the process of local (agri)cultural change, which in turn opens the door to new challenges in approaching research and extension. To build more directly upon farmers' creative capacities, the formal research system will need to find ways of developing technologies that are 'loosely packaged', or specifically designed to allow local producers to use their knowledge, skills and creativity in making the necessary adjustments and adaptations that their particular circumstances require. Similarly, in order to build upon and strengthen local traditions of innovation and information exchange, extension will need to largely redefine its role from serving as *the* communication system, providing *the* answers, to one facilitating the creation and movement of ideas and material through the human landscape. Such new directions, however, are contingent upon the adoption of a more dynamic view of the capacities for change within the local cultural systems, and an appreciation that human creativity is vital to the future functioning of healthy human systems.

8 Findings and future practices

This book has endeavoured to illuminate important aspects of human behaviour involved in the evolution of agricultural production systems. Although the glimpses into the 'black box' of local change provided by this study have left a number of areas only tentatively explored, and in some cases have raised more questions than answers, overall the findings shed enough light on the dynamics of individual and social behaviour that drive agricultural evolution to allow a serious (re)examination of the compatibility between these inherent human qualities and the methods and ultimate objectives of most formal development efforts. The goal of this last chapter is to translate the major findings presented earlier into pragmatic recommendations for improving the effectiveness of rural development initiatives. Following a brief overview of the principal findings from the research, the discussion turns to the practical implications that these findings have for the activities and orientation of rural development organizations serving farmers.

Review of findings

In contrast to the widely held belief that rural communities are incapable of coping with the combined pressures of rapid demographic, environmental and socio-economic change, the experiences in the OHVN over the past several decades depict a very different story. While, as with any group, there are limits to rural people's capacity to cope, the findings presented in earlier chapters (3, 4 and 6) describe how farmers in south-western Mali have managed to keep pace with a doubling of the rural population, a marked decrease in land availability, successive droughts, an overall 20–25 per cent decline in annual rainfall levels and the simultaneous transformation of the national marketing, credit and input supply structures. Overall, during this period of intense change the comparative contributions made by the principal research and extension programmes serving farmers in the OHVN have been minimal (see Chapters 5 and 6). The value of this observation, however, lies not in the basis that it provides for launching further criticisms of the limitations of conventional, top-down R&D strategies. Rather, such a finding highlights the positive aspects of farmers' analytic abilities (however unsophisticated and disjointed) in identifying, and not following, ineffective advice, and the equally important capacity that they have demonstrated in salvaging and modifying useful bits of technology, and in generating their own viable alternatives that have been essential in enabling them to meet the many challenges that they face.

These expressions of human resilience and creativity are first explored in the case-study within the context of the physical environment in which farmers live. As described in Chapter 2, the very foundation of farmers' production systems rests upon a complex and constantly changing mosaic of physical resources. The essential characteristics of these major resource systems (rainfall, soils and genetic material), their diversity and propensity for variation, provide important lessons both for better understanding farmers' behaviour and for identifying the most effective orientation for assistance activities. On the one hand, certain elements of the physical environment, such as soil nutrient levels, are relatively stable, showing moderate, discernible year-to-year changes that are open to management interventions. Other variables, such as annual rainfall (its onset, duration, spatial and temporal distribution) and the associated flood levels, offer less scope for proactive interventions and must largely be anticipated and reacted to within the context of a particular season. Farmers' use and modification of different vegetative communities, responses to changes in weed populations and management of crop genetic diversity represent yet another parameter important in their struggle to achieve greater economic and nutritional security.

Such a diverse and, with regards to the decline in precipitation, increasingly risk-prone environment, places a premium on both farmers' knowledge and ability to (re)act in order to maintain past levels of productivity, let alone achieve any significant increase. In formulating their plans and responding to emergent demands, agricultural decision-makers rely upon their understanding of the major system dynamics (e.g. climate and changes in soil fertility) and the detailed characteristics of specific sites (reflecting both natural conditions and their history of past management decisions), as well as an intimate knowledge of many of the general agro-ecological impacts that alternative management responses will produce. In addition to the inherent vagaries of the physical environment, farmers' decisions must also accommodate an equally variable and uncertain set of household and local social resources—labour availability, debts, loans, human and animal health, state of equipment repair and market prices, among other factors. Having evolved within such a milieu of extreme variability, farmers' management practices are best characterized by their flexibility, adaptiveness and attention to detail. Building upon Richards' (1989b) analogy of agriculture and musical performances, the 'melodic themes' and 'improvisational expressions' important to successful farm management in the Sahel are introduced in Chapter 3, which depicts the major, recurring elements in farmers' plans, practices and their spontaneous adjustments to unforeseen events.

At the household level, decision-makers are seen to defray risk and exploit the available opportunities by diversifying their investment of scarce resources across a wide range of economic activities. These activities include sales of agricultural produce (cash crops and food crop surplus, livestock and animal by-products), value-added processing, as well as local on- and off-farm employment and short- and long-term labour migration. As with farmers' agricultural performances, the exact pattern of investitures pursued in any given year is continually in flux, depending upon the perceived array of opportunities and constraints emanating from both within and outside the household. Similarities in large-scale environmental features, current and historical influences of transportation infrastructure, market conditions and other supportive institutions have all contributed to shaping the opportunity sets of producers across large areas, and allow the characterization of a number of distinct livelihood 'portfolio' areas.

Over time, the specific agricultural management practices and economic activities exploited by farmers in the different portfolio areas have changed significantly. Farmers identified changes that they have made since the 'time of their mothers and fathers' in virtually every aspect of their production systems. These changes, summarized in Chapter 6, include: the introduction of new crops and adoption of new varieties; major changes in equipment use and cultural practices; an increasing focus on soil fertility and water conservation; the development of new agroforestry and horticultural systems; and a general increase in investments in livestock, among others. Many of these changes have been undertaken in direct response to fluctuations in the physical environment (most notably the sharp decline in regional rainfall levels) while others have evolved in response to emerging market opportunities and land pressures associated with rising population levels.

The identified changes in local farming systems have grown out of both direct and indirect contributions from multiple 'sources' of innovation, e.g. public research and extension programmes, private-sector interests, NGO activities and farmers' own initiatives. When viewed together, the types and relative extent of contributions made by the different actors involved reveal certain patterns. Formal R&D efforts, for instance, have had their greatest success in introducing technologies that are based upon materials and approaches originating outside farmers' range of resources and experiences, e.g. inorganic inputs and certain crops and varieties; there has also been some success in refining, re-introducing and promoting certain indigenously developed technologies, e.g. the use of rocklines, living fences, manure, compost and additional crop varieties. Overall, however, farmers' creativity in independently generating innovations and undertaking the needed modifications of technologies developed by formal R&D efforts has played a pivotal role in enabling rural producers to keep pace with the breadth and rate of changes with which they are confronted.

Much of the knowledge, new information and materials supporting farmers' innovative activities are exchanged through informal channels of communication. While many aspects of farmers' general beliefs, world-view and agro-ecological knowledge are broadly shared, farmers' more specific technical knowledge, as well as their individual skill in and rationale for undertaking different actions, is based largely upon personal experiences and is shared on a far more limited basis. Gender, age, kinship ties and ethnicity, among other social factors, were found to exert general influences over each individual's life experiences and access to resources. These same social influences also feature prominently in shaping the principal lines of communication through which individuals exchange information, genetic material and experiences. As depicted in Chapter 4, local knowledge, creativity and the informal processes of information exchange have served as the essential life-blood of local agrarian change, allowing rural households to continually adapt and evolve their practices in response to the changing sets of physical and social conditions, as well as their own desires.

The focus on specific details of the case-study is broadened in Chapter 7 into an examination of the larger patterns of human behaviour important to the evolution of local agricultural systems. Observations on individual behaviour and social interaction are reinterpreted using a synthesis of theoretical concepts and postulates drawn from the vast behavioural sciences literature. This synthesis highlights: the existence and importance of both normative and individual elements of understanding; the influence of different social frames of reference, personal experiences and idiosyncratic differences in the accumulated individual biographies of skills and knowledge; the different states, forms and levels of knowledge; and the essential contributions of human creativity, discovery and informal lines of communication to individual learning and, ultimately, (agri)-cultural evolution. These observations not only help to improve our general understanding of the behavioural processes involved in local change, but they are also instructive in helping to identify important methodological issues and in re-evaluating broader concerns over the underlying objectives and assumptions of development interventions.

Future practice

The major findings from the case-study, in particular the central role played by farmers' knowledge, creativity and communication in local change processes, allow a number of recommendations to be made that hold promise in improving the impact of formal investments targeted at assisting farmers in diverse and risk-prone environments. In order to broaden the relevance that these recommendations may have for areas beyond the context of the OHVN, the following discussion is pitched at a far more general level than has hitherto been the case.[1] Issues are organized along the main themes developed in earlier chapters, and are directed at the formal research and extension (R&E) activities undertaken by public, private and 'non-governmental' organizations.

Farmers' performances and the physical and social environment

Within the context of development efforts in Africa's many difficult production environments, Belloncle's observation that 'if African farmers as a whole have not yet adopted the practices recommended to them, the reason is not that they are too complicated for them; on the contrary, they are too simple to solve farmers' problems' (1989:39) speaks volumes. In the attempt to mediate between the opposing tensions of the very specific needs of individual farmers and the need to generalize in order to make research investments efficient, farming systems R&E programmes have struggled to operationalize the notion of the 'average' farmer, or 'average' farm conditions, within contexts where local farming systems represent diverse and constantly moving 'targets' (Beebe, 1994; Maxwell, 1986). Despite the nature of the demands, R&E programmes, such as those in the OHVN, have shown a tendency to adopt fairly mechanical and unimaginative approaches to confronting agricultural challenges (Tripp, 1992). The result has been the generation of a limited range of static recommendations (e.g. one intercropping pattern, or one fertilizer application rate) extended in blanket fashion, often with little or no regard for important physical and socio-economic differences. In the end, such static efforts have provided little scope for helping farmers to improve the management of their very diverse and dynamic farming systems. One of the preconditions, therefore, for improving the impact of formal R&E activities will be to pay increased attention to the inherent diversity and uncertainty associated with local farming systems. In short, formal R&D programmes will need to recognize what is required in order to be a successful farmer under the prevailing conditions, and to structure their activities accordingly. At a conceptual level, the understanding of agriculture as an evolving set of interlinked performances, introduced by Richards (1989b) and built upon in this study, offers a useful point of departure from which formal R&E initiatives can begin to re-orient some of their field activities. Improvements in this area can be pursued along at least two different, but complementary tracks: raising the quality of individual technologies developed, and expanding the overall range of options offered.[2]

At a minimum, in order to improve the quality of individual technologies, formal R&E efforts will need to reflect a basic compatibility with the important attributes of the major physical-resource systems, as well as the important features of system interaction, with which farmers must contend. For example, new technological options should be subjected to a rigorous sensitivity analysis on their performance under varying amounts of rainfall (an exercise that often is not undertaken). Similarly, the fairly regular patterns of soil diversity, common to many local landscapes and central to farmers' management strategies, should clearly play a prominent role in the conceptualization and organization of research trials, as in the 'toposequence' approach developed in West Africa by ICRISAT in the late 1970s. Such an approach could be further strengthened by using indigenous soil typologies as the basis for organizing trials (cf. Thomasson, 1981)—a practice first used by colonial officers in the Office du Niger nearly 50 years ago. In response to the important interaction between highly variable rainfall (temporally and spatially) falling across diverse local soils, a more sophisticated and holistic view will need to be adopted if efforts to assist farmers in making critical decisions, such as planting, are to be continued. In the case of the OHVN, farmers were nearly unanimous in their appreciation of the planting recommendations provided through the radio broadcasts of the Projet Agro-météologie (which are based upon yield projections using data from over 30 years of field measurements and a network of centrally placed village rain gauges), yet none followed the recommendations because they did not fit their specific field conditions.

If such efforts are to provide farmers with a truly useful resource, a way must be found to link more closely the power of long-term statistical analysis with the individual field-level conditions which farmers manage. One such approach can be envisioned through the development of field techniques that would allow farmers to assess the moisture content of the major soil types that they manage on a field-by-field basis (similar to the hand augers developed in the Australian 'Landcare' movement) in a way that would correspond to the information/recommendations broadcast through the existing radio network. To help build farmers' confidence in such a system, extension field agents could assist farmers in making comparisons between the messages broadcast and the many natural cues, such as plant and bird behaviour, that farmers already utilize.

In addition to improving the quality of individual technological recommendations, formal R&E activities could also do much to broaden the range of different alternatives being offered. This can be pursued in at least two directions. One concerns the development of adaptive, decision-supporting 'tools' (such as the aforementioned example of field techniques for soil-moisture assessment), or the creation of sets of sliding recommendations that can be used by farmers across a range of conditions (e.g. Okali *et al.*, 1994b). To facilitate such efforts, the modelling of farmers' decision trees (see Gladwin, 1989) may be particularly useful in helping to structure recommendations around the key decision points faced by farmers.

The second orientation towards broadening farmers' options concerns the often suggested provision of 'baskets of choices' or 'tool kits' (e.g. Chambers, 1983; Richards, 1985) from which producers can pick and choose. The impact of such 'baskets' or 'kits' could be enhanced by grouping a number of technical options around specific management problems, or key production principles. For example, if the construction of rock barriers effectively improves the infiltration of surface runoff, leading to higher yields and improved food security in years of highly erratic or poor rainfall, and not all locations have sufficient quantities or the means of transporting the large quantities of rock required, then other means of barrier construction could be examined and promoted (e.g. earthen bunds, debris, or vetiver grass hedges) (NRC, 1993; World Bank, 1990). Similarly, as the use of AT becomes increasingly profitable with the rise in rural population densities, yet where not all households own or have access to bullocks, technologies for the use of a wider range of draught animals (e.g. donkeys and horses, which are often more locally important) could be assembled and promoted. The promotion and facilitation of *in situ* conservation and exchange of diverse local seed-types, along with research-generated varieties, is yet another of many areas where such a broadening of choices could be accomplished.

An important conceptual issue underlying both orientations—the provision of 'baskets of choices' and creation of decision-enhancing tools or guidelines—concerns the identification and organization of efforts around the operative 'production principles' that are central to farmers' decisions in mixing, matching and adjusting management techniques. In this regard, the notion of environmental, or production system, 'literacies' (e.g. Maniates, 1993; Orr, 1992; cf. Campbell, 1996) may provide an effective organizing perspective, where a common understanding of the important bio-physical and socio-cultural dynamics can be built between farmers and outsiders through local analysis and joint learning, and used to focus subsequent efforts aimed at exploring various management options. Examples of this type of approach have begun to appear in an increasing number of very diverse and successful R&E field programmes, ranging from soil-fertility management (e.g. Defoer *et al.*, 1996) to integrated pest-management 'Farmer Field Schools' (e.g. ILEIA, 1997) and aquaculture development programmes (e.g. Simpson, 1998). Such an orientation could help to eliminate the artificial separation between natural resource conservation issues and traditional production-oriented themes that exists in many R&E efforts (as was the case in the OHVN, where rocklines were promoted exclusively on their merits of erosion control, and not moisture conservation, and as such were of only moderate interest to farmers). Extension programmes, in particular, must learn to emphasize these central operating principles in organizing their efforts to promote technological options targeting common production problems. A 'production principles' focus could also be used by research in designing suites of 'loosely packaged' companion technologies, e.g. possible changes in cropping patterns near rocklines which are designed to fully exploit the more favourable growing conditions that are produced (i.e. the type of additional changes that farmers begin to explore on their own) (see Atampugre, 1993).

A household livelihood perspective

Among the many operational challenges facing formal R&E programmes is the expanding nature of their mandates. In particular, the increasing focus on alleviating rural poverty, as opposed to a purely production focus, is an emerging area of concern that many agricultural development programmes are poorly prepared to address. Evidence from the case-study suggests that in light of both the nature of the physical environment and the diversity of household resources, no single innovation or change will be capable of stimulating widespread economic growth across all households within an area. Economic growth will continue to come from a diverse set of sources, with benefits accruing unevenly across households (and individual households through time), depending upon their circumstances, objectives and abilities. In fact, it will be important to maintain such diversity, not only for the efficient utilization of diffuse and limited resources, but also in the interest of supporting household security and ultimate survival. If a poverty focus is to be pursued, the pervasive diversity of household production systems must be turned into a positive theme around which specific actions can be organized. To do this, formal R&E initiatives, especially those utilizing a farming systems perspective, will need to redefine and broaden the parameters they use in characterizing research and recommendation domains in order to reflect the variation in household opportunity sets. As suggested in Chapter 3, a characterization based upon the portfolio of economic livelihood activities exploited by households in different areas is one such option. As a planning tool, adopting a portfolio outlook would provide researchers with a conceptual framework for quickly scanning the breadth of principal enterprises upon which households rely in different locations, and identifying points of leverage where existing activities can be improved and new alternatives developed—in effect, using portfolio descriptions as a basis for conducting rapid, sub-sectoral type assessments of each enterprise, across the range of enterprises in which households of different areas are engaged. Such a perspective is reminiscent of those advocated in the early period of the farming-systems research movement, yet one which has seldom been operationalized.

Any discussion of the potential broadening of research mandates must, however, squarely address the dangers of diluting resources and losing research focus—issues that are particularly important for countries with small and/or weak national research systems. In this regard the adoption of a livelihood portfolio perspective offers real potential as an improved planning tool. Not only could such a perspective be used to target more effectively areas of particular concern to farmers, and areas of potential impact for research, but it could also be used as an overall framework in co-ordinating the search for and importation of technologies from other programmes. To help lessen the burden on research institutions that such a broadening in focus implies, the adoption of a portfolio perspective could also prove useful in redefining and redistributing the workload between research and extension programmes. Overall, the under-employment of extension programmes is one of the least appreciated, yet most pervasive problems affecting R&E efforts in Africa. The re-characterization of R&E activities along the lines of existing household livelihood portfolios could enable extension efforts to customize their programming more effectively in order to meet the diverse needs of farmers in specific areas. The decentralization of technical planning in response to the needs of different portfolio areas would provide field agents and subject-matter specialists with a new range of tasks and, in many instances, would begin for the first time to bring their creative capacities into the development process. Extension organizations could also become far more involved in the proactive transfer and refinement of new technical alternatives, which for many technologies may require only minor adjustments in order to capitalize on the potential 'spillover' of technologies generated by other regional development programmes. Such adaptations could be carried out through the comparatively 'light' involvement of subject-matter specialists (or para-researchers) within extension programmes, with some additional technical backstopping from individual researchers, and thereby avoid the delays and drain of resources through 'heavier' investments in on-farm research efforts. Any such move would, however, necessitate a major re-working, and possibly discarding, of the T&V model that has spread throughout much of sub-Saharan Africa.

Formal R&E efforts and farmers' role(s) in development

The final group of recommendations suggested by this research pertains to the re-evaluation of farmers' role(s) in the development process. One of the contributions of this research has been in documenting the reality of the multiple source of innovation process in local technological change, and, in particular, the central role of farmers' contributions. In order for this evidence to lead to any tangible improvements in formal R&E programmes, it will require those involved to re-think some of their basic operating principles. If rural people's knowledge and creative capacities are ever to become anything but 'the single largest... resource not yet mobilized in the development enterprise' (Hatch, 1976), it will be incumbent upon R&E organizations to take the lead in providing the framework within which farmers' potential contributions can have greater impact. One of the central objectives, therefore, of any reformation of formal R&E approaches should be to improve the capacities of the various partners to contribute—farmers and farmer groups, extensionists and researchers. At least three basic pathways can be envisioned through which the resources of the formal and informal efforts can be integrated to improve development initiatives: the incorporation, or direct utilization, of rural people's knowledge and perspectives in problem identification and solution testing; the more collaborative interaction of researchers and farmers, along the lines of one or more of the many participatory R&E approaches; and through supporting and strengthening the informal processes of innovation and communication by providing farmers with new information, materials and opportunities, and then tracking and helping to further refine and spread developments that farmers are able to generate through their own activities (cf. Biggs, 1989b; Waters-Bayer and Farrington, 1993).

The first type of integration, e.g. making use of rural people's extensive knowledge, is an area where surprisingly little has been done on a systematic and broad-scale basis (e.g. Zolad, 1985), despite the rhetoric about 'building on farmers' knowledge and practices' that extends back to the colonial period. Contemporary debates over the non-transferability of local knowledge, and the danger of using indigenous knowledge out of its social context, or the suggestion that little remains to be discovered, lose much of their power when one looks at recent history. Save for the introduction of inorganic inputs (i.e. those components that have proven least useful for farming systems in marginal environments), the vast majority of technologies being promoted in sahelian West Africa, as well as other regions, have their roots in indigenous practices, e.g. animal traction; use of compost, animal manure and nitrogen-fixing trees, shrubs and crops; conservation tillage, rocklines, vetiver grass and terracing; living fences, alley cropping, relay and intercropping; gravity-fed irrigation; indigenous food crops and specific varieties (such as genes for dwarfing and disease and pest resistance); indigenous livestock breeds; natural insecticides, such as neem; and market-orientated home gardens, to name just a few.[3] In addition to the lost opportunities that have resulted from failing to undertake a more systematic assessment of indigenous practices, other opportunities have been lost in terms of local confidence building, the ability to stimulate further innovation among farmers and the synergy that might have resulted from closer interaction between formal R&E programmes and farmers.[4]

One theme that will be important in improving the level of understanding of farmers' practices and knowledge is that of knowledge differentiation. As this research has reaffirmed, farmers' knowledge and practices vary widely, both within and between communities. Such differentiation not only reflects the specific demands of local conditions, but also the inherent quality of knowledge formation, i.e. the evolution of agricultural knowledge and technologies does not proceed at a uniform rate across locations, even when basic conditions are similar. Building upon the notions of constructing environmental or production system literacies and assembling corresponding bundles of options for farmers, as discussed earlier, one area that holds particular promise is the 'smoothing out' of differences in the existing knowledge base among farmers in different locations and R&E programmes. The development of expert information systems, based upon the combination of the best of both local and formal knowledge (e.g. Guillet *et al.*, 1995; Walker *et al.*, 1991; cf. Thapa *et al.*, 1995) is one way that this can be done. Although technologies involved in this type of approach are still in their early stages of development, they have the potential to provide the added benefits of creating a concrete framework for integrating formal and informal knowledge and building a shared vocabulary and understanding (i.e. literacies), as well as helping to establish a focal point for sustaining the productive working relationships between farmers, researchers and extension field agents.

The second, and perhaps most problematic, type of integration concerns that of farmer–researcher collaboration. Examples drawn from agriculture, research and industry (e.g. Gamser, 1988; Sperling *et al.*, 1993; von Hippel, 1978; amongst others) show that if the operational challenges, and in the case of agricultural R&E the often debilitating biases against farmers' capacities (e.g. Bentley, 1994; Chambers, 1983), can be overcome, the potential gains are substantial. In Mali, some of the most successful research initiatives undertaken by the different national field research units illustrate the significant benefits that the involvement of technology users in the development process can bring (see Kingsbury *et al.*, 1994; McCorkle *et al.*, 1993). First, however, a number of institutional barriers will have to be removed. At a minimum, the internal culture and organizational incentive structure (financial and technical support, professional recognition and promotion) will need to be compatible with any new goals. Where organizational changes are necessary, it will be important to keep in mind that just as farmers' conditions differ widely, so do those of individual R&E institutions. The dogmatic implementation of static participatory approaches will be less effective than a more flexible examination of different ways in which basic principles can be put into action. While, ultimately, collaborative farmer–researcher interaction has the potential to create a cultural 'free space' where innovation is encouraged, and even expected, much ground will need to be covered before such potentials can begin to be realized.

A certain degree of overlap exists between farmer–research collaboration and the final form of integration involving the 'feeding' of the informal processes of innovation and communication. One key difference lies in the primary intent, which for the latter is the explicit strengthening of farmers' abilities to explore new options and exchange experiences and information. The rationale for promoting this type of integration is found in the many, albeit entirely unplanned, examples of farmer-driven technological change described in Chapter 6, and which are the subject of a rapidly growing body of literature on indigenous experimentation and communication (e.g. Alders *et al.*, 1993; Haverkort *et al.*, 1991; Hiemstra *et al.*, 1992; Prain *et al.*, forthcoming; Sumberg and Okali, 1997; van Veldhuizen *et al.*, 1997). As demonstrated in this research, the flexible use of rapid research methodologies is one way to monitor farmers' innovative activities with a minimal investment of resources, a process which could be greatly enhanced by the existence of local organizational structures.

Perhaps the greatest challenge, at least initially, in placing a greater emphasis on any of these approaches (drawing upon, working with or contributing to farmers' own efforts) lies in building up the individual and organizational experience necessary to effectively integrate such alternatives into existing planning and operational procedures, as well as understanding when and where these procedures themselves must be changed. The success of institutionalizing such efforts will rely not only upon the ability to make good decisions a priori, but also in introducing more open and flexible decision-making structures that allow the best path to be identified as one goes along.

For extension efforts in particular, these three types of interaction—drawing from, collaborating with and feeding local processes—have far-reaching implications. One area that will be increasingly important is the strengthening of local organizations and social networks. Experience has shown that in order for research(ers) to work effectively with farmers, some type of local organizational structure and common language is needed. Common sense, and the limited evidence available (e.g. Romanoff, 1990; cf. Bebbington *et al.*, 1994), suggests that it may be prohibitively 'expensive' for researchers to become involved in this sort of preparatory activity. Nor are researchers trained or organizationally best suited to undertake such activities. This is not true for extensionists, however. Although most extension (and NGO) programmes operating in Africa already rely upon some type of 'group' approach, few involve groups in active research and dissemination. Nevertheless, a growing number of examples showing the successful use of 'group' processes in developing and spreading new technologies can be found in R&E programmes that cut across the African continent.[5] An increased focus on expanding the use of group-based activities would provide yet another area for the meaningful expansion of the field-agent's role, and represents a concern for human capacity-building that has been absent from most extension programmes (cf. Röling, 1986).

Beyond the generation of new technologies, another fertile area for extension programmes concerns the broadening of the range of approaches used in exchanging information and experiences among producers. Mundy and Compton (1991; 1995), for example, have developed a simple model (Figure 8.1) that could easily be transformed into a heuristic planning device by any extension programme to aid in optimizing the dissemination of different types of information. In addition to the existing focus on extension meetings and demonstrations, the findings from this research in the OHVN (e.g. the importance of travel and structured exchange visits with other farmers) suggest that a combination of approaches that help to stimulate farmers' curiosity and broaden their range of experiences will be critical. A vastly increased use of farmer-to-farmer visits, regularly scheduled farmers' radio programmes, the creation of audio/video tapes of farmers' experiences with various technologies, work with innovator 'clubs' and the development of farmers' best ideas into technical recommendations offer a number of possibilities for helping to revitalize the impact of extension efforts. In addition, specific 'nodes' in the local communication pathways, once identified, could be used to facilitate the diffusion of information. The operation of simple information booths at key weekly markets, and the use of existing lineage, age-set and traditional work groups as the basis for communicating information are just a few possibilities.

A deeper, and less obvious, theme that runs through many of the preceding recommendations on the increased involvement of farmers in formal R&E activities concerns one of the chief findings elaborated upon in Chapter 7—that the patterns of human behaviour observed in the OHVN are in fact those responsible for (agri)cultural evolution and survival (cf. de Schlippe, 1956; Johnson, 1972; Loevinsohn and Simpson, 1998). Clearly, one of the important findings to emerge from this research is that the greatest potential for stimulating and sustaining rural development may lie in the ability to encourage and strengthen the internal

		Types of knowledge	
		Exogenous knowledge	Indigenous knowledge
Types of communication channels	Exogenous communication	Technology transfer	Indigenous knowledge-based Development
	Indigenous communication	Diffusion; Co-opting of folk media	Cultural continuity and change

Figure 8.1 *Communication channels and types of knowledge* (*Source*: *Mundy and Compton, 1991*).

processes of socio-cultural change within local communities themselves. Nor is such a perspective new. Earlier in this century, Gandhi, among others (e.g. Nyerere, 1968), regarded rural development in terms of the moral and intellectual development of the individual, where the development of individual creative intelligence, won through the identification and solution of problems of immediate relevance, was seen as the central engine of rural development (Herrera, 1981:21). While such a view is supported by much of the empirical evidence from the case-study, it is almost completely at odds with the service-delivery philosophy currently driving most formal development efforts. Certainly in the interest of improved effectiveness, efficiency and concerns for longer-term sustainability, a larger share of development efforts should be focused upon developing rural capacities through engaging rural people in active, real-life problem-solving efforts. In the end, this may not only contribute to 'better science,' as noted by Richards (1989a), but also to a more sustainable form of rural development.

A final thought

Owing to the conditions under which this research was carried out, certain time-dependent features of farmers' behaviours, such as the full range of intimate details of individual farmer's within-year adaptive performances and the year-to-year changes in household investment patterns, could not be thoroughly explored. The point, however, has long been passed where more research is needed on farmers' knowledge and capacities for the sake of advocacy. The major challenges now facing R&E organizations are how to put into practice effectively that which is already known, and how to change the established institutions and education systems in order to facilitate this. On both of these fronts—applying principles in practice and facilitating institutional change—the basic tenets of action research hold particular promise. With regard to the first theme, the American social psychologist Kurt Lewin, credited with developing the approach and early practice of action research, advocated the need to 'consider action, research and training as a triangle that should be kept together for the sake of any of its corners' (1946:42).[6] The conceptualization of 'action' as the use of theory in practice, the goal of 'research' as contributing to theoretical enrichment, and 'training' as the acquisition of capacities to undertake future actions and research, offers a useful paradigm for any organization concerned with rural development.

A second theme in the action research tradition that holds particular promise for improving agricultural and natural resource management initiatives concerns that of organizational development, e.g. enabling development organizations to undertake the internal changes necessary to carry out their missions more effectively (e.g. Argyris *et al.*, 1985; Foster, 1972; Whyte, 1991; amongst others; cf. Selener, 1997). Similar to the adaptability that farmers have shown in their agricultural and economic performances, formal R&E programmes must provide for their own learning and evolution. At present, however, few such programmes possess the characteristics of true learning organizations (see Senge, 1990). Possibly one of the greatest lessons that R&E organizations can gain from farmers is the ability to 'embrace error' and learn from their mistakes, without losing the confidence and drive to undertake new efforts. Based upon the findings from this research, the over-arching need to unleash the human potential (in terms of knowledge, creativity and communication) of farmers, field workers and researchers is an objective that should lie at the heart of any future development initiatives.

APPENDIX A
Research methods

At the risk of creating the illusion of a problem-free and smoothly implemented research design, many of the twists and turns common to any field research experience have been eliminated to avoid burdening readers with unnecessary details. Only those aspects of the research that had significant influence on the outcome or which served as important guide posts in the process are included here (see Simpson (1995) for a more detailed description of research methods).

Data collection and analysis techniques

In addition to the usual influences of time and financial constraints, the large size and diverse qualities of the research area (as well as the exploratory nature of the central research questions) argued for the use of highly flexible and adaptive data collection and analytic techniques. The overall strategy of the Grounded Theory approach (described in Chapter 1) was therefore blended with various Rapid Rural Appraisal (RRA) data-collection methods.[1] Structurally, the study was organized around a multi-stage research design that allowed a progressive immersion into the physical and social contexts of the study area. These stages included: a period of pre-departure data collection, review, and preliminary problematization of key research issues; a rapid reconnaissance of the study area involving forays into the field, and interviews with various individuals associated with public, private, non-profit and volunteer organizations active in the study area; the main phase of data collection and initial rounds of data analysis; and successive rounds of increasingly intensive post-field data analysis and integration of secondary information.

Research preparations

A two-phase plan for the field research was devised, based upon information gleaned from the project literature (project documents, annual reports and commissioned studies), and interviews with persons familiar with the study area and OHVN organization. The first phase was designed to provide an initial three-week reconnaissance of the study area where the main research questions, assumptions and proposed data collection techniques could be tested, refined and, if need be, modified or replaced, followed by the second (main) phase of data collection and primary data analysis.

Drawing upon general approaches to rapid data collection (e.g. Bernsten *et al.*, 1980; Holtzman, 1986), a four-step process was used in preparing for the initial reconnaissance of the study area. First, the major areas of inquiry were broken down into individual research issues and questions. For each question, a minimum data set was established, i.e. the minimum data thought necessary to begin addressing that particular question or issue. Second, based upon the interviews and review of literature, each of the identified data needs were linked with at least two potential 'sources' thought capable of providing the needed information, as well as at least one additional source that could help to verify these primary responses. Third, the data needs and potential sources of information were then paired with specific data-collection techniques deemed appropriate for the particular source and the type of information required, and which were manageable under the time constraints. Overall, the selection of different sources of information and data-collection methods was guided by the principle of 'triangulation,' a central feature of the Grounded Theory approach (Glaser and Strauss, 1967), and one which has become critical to the success of RRA fieldwork.[2] Finally, after linking the basic information needs, potential sources and the use of different methodologies, a tentative schedule was drawn up which provided an opportunity to access each of the sources identified during the reconnaissance phase. Although these pre-field work preparations were important in providing the research team with its initial orientation, they were undertaken with the expressed intent that they would be extensively modified during the reconnaissance itself, as well as during the preparation for the main phase of field research, as additional insights and information became available. To facilitate this in situ modification of the research design, questions and methods, a 'toolbox' of research materials was assembled and taken to the field.[3]

Rapid reconnaissance

Following an initial round of introductions and interviews with the programme staff, donor representatives and project personnel from the principal governmental, non-governmental, private-sector, bilateral and international organizations supporting rural development activities in the OHVN zone, a revised programme of visits was drafted that allowed the research team to visit 11 villages over an 11-day period (see Figure A.1).

During these village visits, group and individual semi-structured interviews, various diagramming techniques (e.g. Venn diagrams of social organizational interactions; preference ranking of various 'sources' of new materials and technologies) and field visits were used to gain a sense of the local development dynamics. The information generated in these visits led to a number of significant changes in the proposed design of the main phase of the field research, the most significant being the decision to eliminate the use of village-level and individual questionnaires. The use of lengthy questionnaires involving large sample populations had become the favourite tool of the OHVN and other governmental agencies in collecting field data in the area. In many cases villagers had been subjected to several surveys each year for the past several years, leading one farmer to remark that 'they ask us the same questions every time, but never do anything'. Drawing upon the strengths of the most successful methods used during the reconnaissance visits, a sequence of group and individual semi-structured interviews, preference ranking, pie diagrams and resource- and activity-mapping exercises, guided field visits and direct observation were proposed as the primary research techniques for the main phase of data collection. This complementary set of research activities, and a completely revised set of research questions, were then field-tested and further refined during an all-day visit to a village 20km outside of the capital.

In addition to guiding these methodological changes, insights gained from the reconnaissance visits also led to other important changes in the overall conceptualization of the research. The initial village visits revealed that the OHVN extension service was without question the principal programme involved in the promotion of 'new' agricultural technologies to farmers across the zone. Even so, the assistance being provided by the OHVN, and its partner organizations

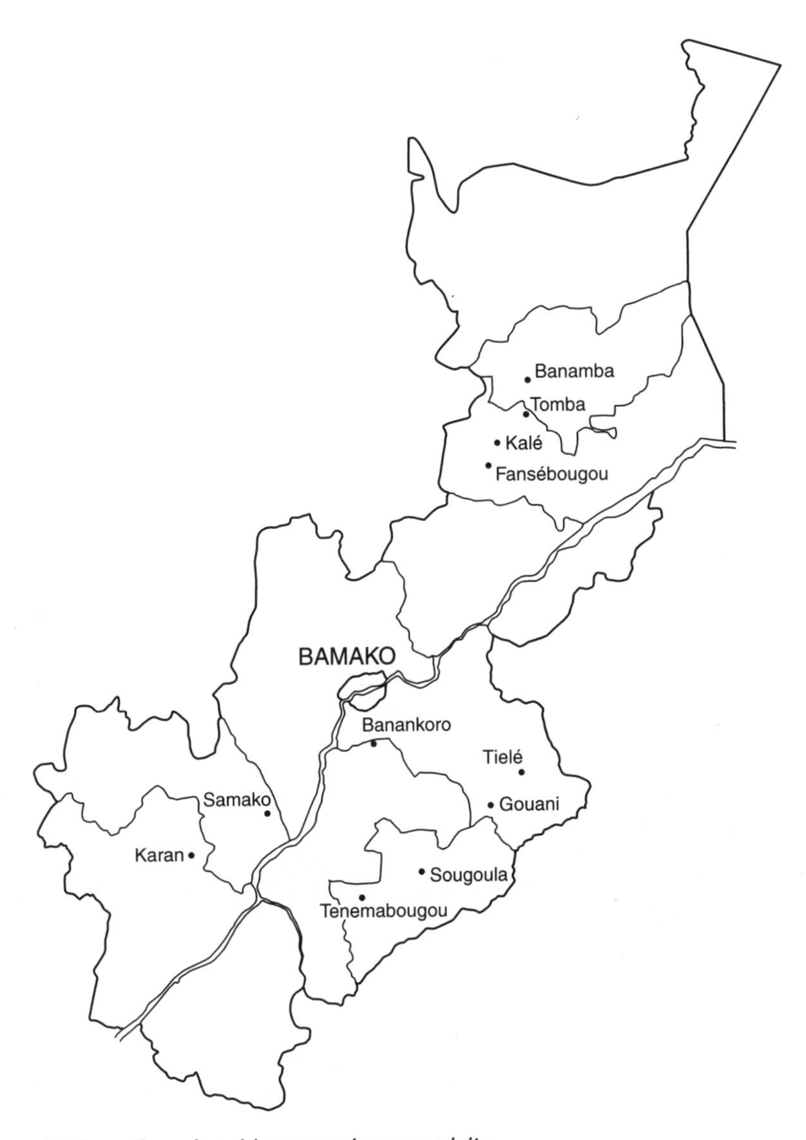

Figure A.1 *Location of rapid reconnaissance visits*

(CLUSA, DRSPR/OHV, FAO, PAE, Peace Corps—see Chapter 5), was quite limited, and in most cases stopped literally at the village edge, i.e. there appeared to be little outward spread of its impact. Much of the limited diffusion of the programme's messages could be traced to the integrated 'package' approach used by the OHVN, where the principal technologies being promoted were highly dependent upon the provision of credit, production inputs and marketing services, which the OHVN had also historically supplied. In contrast, the more dynamic process of indigenous development seemed to permeate the entire area, seemingly little influenced by the extent of contact with the formal system. Thus, in terms of sampling, the decision was made to concentrate on examining in greater detail some of the OHVN's 'best case' villages, in order to gain a clearer picture of the OHVN's potential for stimulating local agricultural development throughout the project zone using its current approaches. At the same time, comparative observations could be made on the contributions of the parallel processes of indigenous innovation, adaptation and communication.

Main phase of field research

Site selection Based upon the findings of the reconnaissance, three criteria were used in selecting sites for the main phase of field research: agro-ecological conditions (principally annual rainfall levels and soil types); the institutional history of an area, in terms of its past relationship with governmental development programmes; and the intensity of current development activity by the formal research and extension systems. With respect to the physical conditions, the main agricultural area of the OHVN zone was divided roughly in half, with the northern portion of the zone consisting of areas with predominantly sandy soils and averaging less than 800mm of rainfall a year, and the southern areas situated primarily on granitic soils, receiving up to 1200mm of rainfall annually (see Chapter 2). An examination of the institutional history of the OHVN zone revealed that the northern areas of the zone had once been part of the Opération Arachide et Cultures Vivrières (OACV), the Malian parastatal responsible for promoting peanut production, which was transferred to the OHVN in the early 1980s. Areas of the south-east had formerly been part of Compagnie Malienne pour le Développement des Textiles (CMDT), the cotton parastatal, which had been transferred to the OHVN in the late 1970s. The history of development efforts in the central and south-western areas of the Haute Vallée dated back to the mid-1960s, when tobacco, rice and vegetable crop production initiatives had been strongly promoted (see Chapter 5). With respect to the intensity of formal development activities, the third criteria, the consideration of possible study sites was limited to those villages with an established village extension group. In addition, those villages participating in the Département de Recherche sur les Systèmes de Production Rurale (DRSPR) field research programme, the FAO decentralized seed-multiplication programme and additional non-governmental and bilateral development projects were especially targeted.

Based upon the first two criteria—agro-ecological conditions and institutional history—the OHVN zone was divided into three areas: the north, south-west and south-east. In consideration of the third criterion—level of development intensity—three villages were selected in the north (Kanika; Tomba; Fansébougou), four villages in the south-east, (Bassian; Sikoro; Falan; Tinguelé), and four villages in the south-west (Samako; Missira; Déguela; Balanzan)(see Figure A.2). In all, villages were selected from seven of the ten OHVN administrative secteurs.

The principal assumption of the research design—that looking at villages with the highest level of contact with the formal system of research and extension would provide a basis for a 'best case' comparison of the impact of both formal and informal (farmers' own) development activities—led to a heavy emphasis being placed on the third criteria of intensity of formal system contact. As a result the villages selected included four out of five of the DRSPR/OHV main on-farm research villages, and three out of six of the FAO seed-multiplication centres located in the zone. Furthermore, three of the villages housed OHVN field-agents, and another three housed agents of related OHVN programmes.[4]

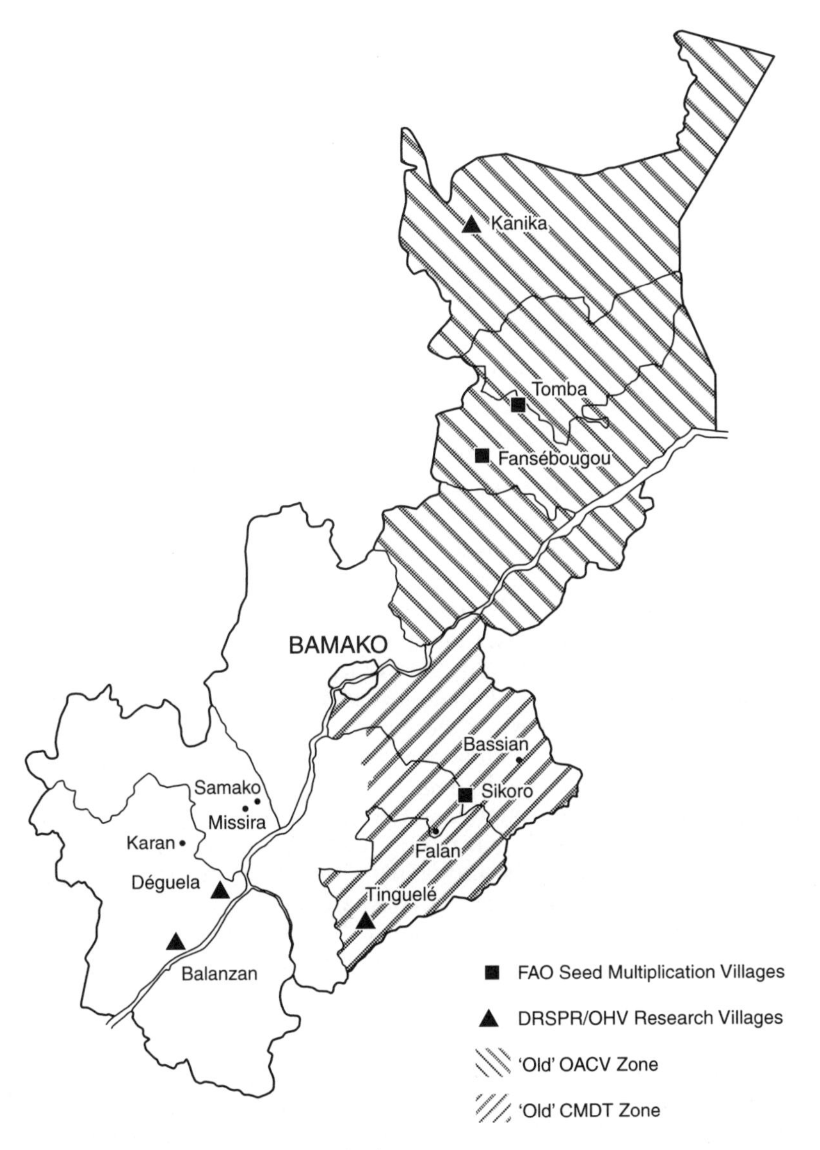

Figure A.2 *Location of main research sites*

All 11 villages were within 25km of the OHVN secteur field offices. The only village included in the study that did not house an OHVN or other field agent, or host a supplemental development activity (e.g. on-farm research programme, or a seed multiplication centre), had recently hosted one of the annual farmer field days sponsored by the OHVN, and had participated in a major varietal/fertilizer field trial jointly sponsored by a private fertilizer company and several governmental research departments.

Village visits Following the selection of study sites, a schedule of visits was established, allowing two or three days of fieldwork in each village. In addition, a questionnaire was administered to the entire OHVN extension staff in each secteur. The main research activities were structured around group and individual contacts, with both men and women farmers focusing on the identification of, and exploring the processes involved in, past and current changes to: agricultural practices; patterns of natural-resource use; major agricultural problems; the importance of various nutritional and economic activities; and the history of village relations with external organizations. In exploring each of these themes, data-collection methods were selected to help reveal the differences between individual and group perceptions, and 'normal' versus actual practices. To take full advantage of the progressive learning offered by each village visit, short periods away from the field were build into the schedule to provide opportunities to review field-notes, reflect on observations, revise questions, raise new lines of inquiry and begin preliminary data analysis.

To facilitate the village interviews, secondary-school leavers were hired as interpreters. In the attempt to avoid possible biases related to project influences and/or local interpersonal relationships, efforts were made to select individuals who were not associated with the OHVN, including village animateurs (who worked for, and were sometimes paid by, the local *associations villageoises*), and who did not reside in the village where the interviews were taking place. These conditions were violated in three cases, yet in only one instance did the local hire of an interpreter result in any serious problems during the interview sessions.[5]

Although all of the interpreters hired were male, this did not noticeably interfere with the interview sessions with the women. As in several of the men's group sessions, local women fluent in French served as the *de facto* interpreters for the interviews in three villages. While the inclusion of these additional participants did lead to some moments of initial chaos in the group sessions, in the long run having more local people involved in the translation process helped to relax participants and invariably improved the accuracy of the translations. While the interviews would possibly have gone more smoothly if a single interpreter had been used throughout, the difficulties that arose proved, ironically, to be immensely valuable learning experiences in terms of providing glimpses into the fragile nature of the relationships between the *associations villageoises* and the unpaid animateurs, and the types of dissonance that can occur in the local social fabric.

Data Collection Data collection in each of the village visits followed a similar though continually evolving pattern. Representatives of the men's *group de vulgarisation* (GV) and the president of the women's association were contacted to arrange for the initial group meetings. The first group sessions were held with the men's GV, while the meetings with the women's groups typically took place on the second day. Attendance at these meetings varied from 12 to nearly 30 members, with participants ranging from their late teens to the very aged. Although men's meetings were intended to be held with GV members, in most cases there was a large number, even a majority, of individuals from outside of the GVs.[6] In contrast, the meetings with the self-formed women's village associations consisted almost exclusively of current members.

At the start of each of the group sessions the intent of the study and different activities were explained. In the initial group sessions, guidelines were used to orientate discussions around farmers' perceptions of change in their agricultural practices/systems over three general time-frames: the period since the preceding generation; the recent past (last three to five years); and the present, i.e. what changes farmers were currently making. Using an approach developed by World Neighbours (Gubbels, 1988), each of the initial group sessions began with a general, open-ended question, e.g. 'describe the agricultural practices used during the time of your mothers or fathers', followed by a series of closed and open-ended questions that began to specify what field types, intercropping patterns, rotations, varieties, cultural practices, equipment, livestock and agroforestry associations farmers used. Next, an equally detailed description was then generated of the current agricultural systems. The differences between the past and present practices were then explored, relying heavily upon the 'six helpers'—what, who, where, when, how and why—to probe into new areas that emerged from the group's responses. In cases where the identified changes involved a specific technology, the creation, source and pathway of diffusion among the community was discussed. Over the course of these initial sessions, attention was focused on increasingly contemporary experiences and processes of experimentation and communication.

In meetings where a 'spokesperson' began to dominate the discussion, the individual was thanked for her or his valuable contributions, and those present were reminded that in this 'group' meeting, everyone's opinions and experiences were needed. It was made clear that the groups did not need to come to a consensus, but that it was important to discuss the different points of view. In every case, once the floor had been opened lively debate and widespread discussion followed.

During the interviews, observations on participants' behaviour served as triggers for further probing an issue, as well as cues for when to move to a new topic, or close the discussion. In this way, each of the group sessions was customized to fit a group's rate of progress. The principles of verification, continual cross-checking of information, maintaining an optimal level of ignorance (not assuming knowledge), reporting back understanding and probing into new and increasingly detailed areas all played an extremely important role in managing these and subsequent interview sessions.

In addition to the initial semi-structured interviews, the group sessions were also used to complete various ranking and diagramming exercises. Not all of these exercises were completed in a single session. Often breaks were taken for meals or prayer, or themes were covered on different days. Exercises were used to first identify and then rank the relative value of different sources of agricultural information, the severity of agricultural problems (past and present) and the relative nutritional and economic importance of different crops/varieties in 'good' as opposed to 'bad' seasons, past and present, as well as additional non-farm, non-agricultural activities. Pie diagrams were used to estimate the relative levels of use of animal traction and the importance of different crops. Using a blackboard, borrowed either from the village school or literacy centre, groups were asked to construct 'maps' indicating the periphery of the village (or different concessions), location of the major roads and pathways, and to use whatever symbols they wanted in representing the location of specific resources important to their livelihoods, as well as specific activities, such as: formal research and informal varietal test plots; fields using technologies adopted from the extension programme, or where new or unique local varieties had been planted; unique production niches; the fields of particular individuals; and any other item of interest that had come up during the interviews, or which the participants wanted to highlight. In about half of the cases, participants needed help with the initial orientation of the map (e.g. locating a few main reference points, such as the main road or stream). In addition to serving as a general 'stock taking' and centre of discussion regarding specific resources, activities or events, these maps were also used as the basis for organizing 'tours' to visit the sites indicated on the maps.

While the group sessions and mapping exercises provided a watershed of information, largely indicative of the 'normative' or contrasting major views on general trends and practices, the individual contacts made during the field visits provided the richest period of more detailed data collection. Following the route laid out at the end of the mapping exercise, the field tours were typically held on the second day of each village visit and lasted between three and four hours, or more, depending upon the number of contacts made. Individuals who had been identified in the course of group discussions as being particularly knowledgeable, or who held obviously different views about various subjects, were often asked to serve as guides for the field tours, or were asked to identify a specific time and location where they could be met during the 'tour' in order to discuss their activities. The field visits were used to compare and contrast information obtained during the group sessions with individual practices, as well as to conduct detailed discussions on: individual field histories; personal biographies with regards to specific crops/technologies; who received and shared information with whom, when and where; and what specific changes farmers were currently making and where the ideas for those changes had come from (cf. Okali *et al.* 1994a; Sumberg and Okali, 1997). These tours provided an especially important opportunity to meet and talk with a wide range of individuals who had not participated in the group meetings.

Following the field tours, a second meeting was generally held with each group. This provided the opportunity to ask questions concerning the many 'discoveries' made during the course of the field visits, as well as a chance to explore the nature of any discrepancies between the information obtained during the previous group session and the information gained from individual interviews, or in discussions with the other (women's or men's) group. These final sessions were the last formal opportunity to collect missing information, or to clarify issues that remained unclear from the previous discussions.[7]

Mid-course adjustments Certain conditions encountered during the course of data collection necessitated modifications to the research design. Initially, equal time had been planned for replicating the full set of interviews, mapping exercises and field tours with both the men's and women's groups. However, because the data collection was carried out during periods of peak labour demand—towards the end of the planting season in the north, and at the time of the first weeding in some areas of the south—it was felt that the additional pressure placed upon the women to participate in these activities was unreasonable. As a result, several adjustments were made to ease the burden, while still attempting to maintain a high level of quality in the information obtained. The number and length of the women's group sessions were limited. Women participants were asked to select the time(s) and place(s) for their meeting that was the most convenient for the majority of their association members, which they greatly appreciated. After the first several village visits, women were not asked to construct their own maps (often a very time-consuming exercise, requiring upwards of an hour to complete), but a representative of the men's group would present and explain their map to the women who, in turn, were given the chance to add, modify or delete items from the men's map (a hand-sketched and photographic record was made of each group's rendition). Although the women were generally hesitant to make major structural modifications to the men's map, they did typically add several new items that were of particular significance to them, such as their personal fields and gardens, group fields, the location of important niche environments and various development projects with which they were involved. Finally, in response to time pressures, in about half of the villages women were asked to identify a time when I could meet with them at specified locations during the field tours, rather than conduct entirely separate tours.

Although on the whole these changes reduced the amount of formal contact time with women, the overall high quality of the meetings with the women's groups and the addition of supplemental activities helped to compensate for this shortcoming. In general, women participants answered questions more readily and came to a consensus, or alternative positions, faster than did the men. Women were less hesitant to voice their criticisms of various aspects of the development organizations with which they were in contact, or of the resource constraints and labour demands that were being placed upon them by their husbands. Information gained from the men's discussion was typically used as a foil against which the women could contrast their own experiences (e.g., 'the men claimed that... ; how does this compare with your experiences'). This greatly accelerated the interview process, and also helped in data analysis and category formation, since the information obtained in this manner was already in a form that spoke directly to the influences of gender on resource allocation, agricultural practices and, in some instances, general knowledge. In addition, the time spent in-between formal research activities was used in individual conversations with women regarding various aspects of their domestic chores and activities. In this way, a wealth of detailed information on women's domestic work, on-farm processing and income-generating activities was acquired, which would have been otherwise unavailable.

An additional modification in the research approach that was made early on in the research process concerned one of the conceptual categories identified in the literature—the notion of local 'experts' (e.g. Box, 1988; McCorkle *et al.*, 1988). After initial difficulties in getting village groups to identify local 'experts,' or those farmers most knowledgeable about specific activities, discussions with other field researchers[8] indicated that farmers may have been avoiding such questions because of the risk of arousing jealousies through publicly emphasizing the differences in individual skill levels. Consequently, the questions were modified to determine from whom individuals had received 'good advice' about producing one or another crop, or to whom farmers had, or would, go for 'good advice,' with the assumption being that these individuals possessed superior knowledge or wisdom concerning specific areas of crop production. This revised approach worked far better in eliciting responses during subsequent village visits, and helped to identify a number of individuals who were highly involved in experimental activities.

Additional activities In conjunction with the village visits, a questionnaire was administered to the entire OHVN field staff. The questionnaire contained six parts; the data from three of these areas—extension-research and extension-farmer relations, and agents' impressions of the extension programme—are included in Chapter 5.

In addition to this questionnaire, informal market surveys were conducted in a number of major market centres (Bamako, Banamba, Bancoumana, Oueléssébougou and Tielé). These surveys helped to confirm the availability and level of private-sector promotion of various types of equipment and inputs, and the regional differences in the economic importance of various crops, gathered products and goods manufactured through on-farm processing. The extensive field travel also provided numerous opportunities for informal discussions with OHVN field staff, DRSPR/OHV researchers and those associated with other development programmes operating in the area. These discussions were immensely helpful in gaining additional perspectives on various aspects of local development activities.

During the often long periods of free time between structured activities, a range of informal discussions were held with various individuals and groups (generally age-set groups). In a few instances spontaneous sessions were held with village 'elders,' about the 'old days' and how things had changed with respect to the present. The opportunity to observe life in the compound of the host family in each village extensively, and talk with family members, especially the women, provided an additional wealth of information.

Data analysis

Consistent with the Grounded Theory approach, the collection and initial analysis of data (the comparison of data for patterns of similarity and difference) took place in an iterative fashion throughout the main phase of the field research. In addition to the informal discussions and opportunities for direct observation, the periods between planned research activities in each village were used to review and expand upon interview notes, record additional comments and observations and begin to look for areas of missing data or issues that would require further investigation. The recommendation by Glaser and Strauss (1967), and ethnographers (e.g. Hammersley and Atkinson, 1983), that researchers should block out regular periods of time away from the field to review and begin to organize their field notes in order to help direct subsequent sessions of data collection proved invaluable; the weekly returns to the capital were used as an opportunity to compare new data with previously collected responses and to search for and challenge the emergence of patterns in the data. As described, the results of these initial rounds of analysis were invaluable in helping to direct and modify the lines of inquiry in the group and individual interviews in subsequent village visits. It was in this way that several of the conceptual categories developed in this study were first 'discovered.' Unfortunately, due to the lack of opportunity to revisit specific sites later in the research process as had been planned, I was unable to use the Grounded Theory approach to its full advantage (i.e. formulating, testing, revising and retesting hypotheses governing the relationships between the different conceptual categories). As a result, much of the theoretical interpretation of the research findings took place in subsequent periods of post-data collection analysis and integration of secondary data.

During the iterative rounds of data analysis, facilitated in the early stages by the reporting requirement stipulated in the Terms of Reference, information from an increasingly broad range of sources was integrated into the analysis. One important stage of analysis concerned the comparison of changes reported by farmers in their production systems with the record of DRSPR/OHV on-farm research trials and extension messages that had been issued by the OHVN extension programme (see Chapters 5 and 6). In other instances conceptual interpretations, such as Richards' (1989b) agricultural 'performances,' were taken directly from the literature and expanded upon. The use, and in the case of the concept of local 'experts,' the rejection, of such concepts is based purely upon the 'goodness-of-fit' of their explanatory powers over patterns of behaviour observed in the OHVN case-study, and is entirely consistent with the Grounded Theory approach, so long as their use is critically reviewed (Strauss and Corbin, 1990). Explicit in the use of all such secondary literature is the direct relationship of the information to the categories generated from the primary data collected during the field portion of this research. These additional sources of data are used to exemplify, explicitly extend, contrast and add further detail to the field observations in helping to generalize these interpretations beyond the limits of the OHVN project boundary. In terms of theoretical development, this is most evident in the material presented in Chapter 7, which reinterprets the patterns of behaviour and relationships identified in the case-study through a synthesis of theoretical contributions drawn from a number of behavioral science disciplines.

APPENDIX B
The emergence of an integrated approach to rural development in Mali

With the election of the socialist candidate Modibo Keita at Independence in 1960, there was little surprise that Mali's first five-year national development plan articulated a form of 'rural socialism' (Bingen, 1985; USRDA, n.d.), termed by one observer as a 'conciliation between Marxism, Islam and traditional communalism' (Megahed, 1970). Much of this plan was oriented towards decolonialization, achieving economic independence and arousing a sense of political consciousness among the rural poor. Yet, there can be no question that the new Malian government saw agricultural and rural social development as the primary vehicle for achieving its political objectives. Through a series of initiatives the government sought to create a 'new man in the countryside' (Bingen, 1985), aided by the establishment of farmer co-operatives, collective fields, local and regional agricultural schools, all mixed with local party representation. '[I]nstead of pursuing economic and social policies to transform peasant... production into socialist production units such as state plantation, the Keita regime believed the state had only to revitalize the productive forces of the traditional village economy' (Bingen, 1985:16–17). The problem lay in the fact that there were not enough trained civil servants and financial resources to fully elaborate and carry out these plans. As a consequence, few of these initiatives ever achieved widespread use (Jones, 1976; Megahed, 1970), and the Malian government was largely forced 'to make do with a continuation of French colonial policies for rural development' (Jones, 1976: 268).

The technological basis

Aside from the creation of Office du Niger in 1932, however, the French had never invested heavily in Mali's rural development. Of those activities that were undertaken, most can be tied to two important technical 'packages' developed by the agronomist Viguier. These 'packages,' aimed at increasing the productivity of traditional agriculture, were to have a lasting effect upon Mali's early agricultural development. Based upon research begun in the 1930s, one of the packages involved the construction of 'polders', locally managed water-control systems designed to gain partial control over the rise and fall of annual flood waters, necessary to improving rice production (Gallais, 1967; Viguier, 1939). The second technical package, also originating from work initiated by Viguier in the 1930s, was oriented towards the production of upland crops involving a specific rotation of cotton (fertilized with manure)–sorghum–peanuts–fallow (using *Stylostanthes gracilis*) (Jones, 1976). This system of upland crop production was referred to as the *M'Pesoba* method, after the agricultural school which Viguier established and where the system was first perfected (Viguier, 1952). Post-Independent agricultural development programmes were to continue to rely heavily upon both specific elements of the polder and *M'Pesoba* systems developed by Viguier, and the more general concept of developing and promoting commodity-based technical 'packages' (e.g. early development efforts in the Opération Haute Vallée included both the promotion of the *M'Pesoba* system[1] and construction of a polder, which continued to receive development attention well into the 1980s (USAID, 1978).

The institutional framework

By the mid-1960s a number of separate, commodity-based, agricultural *programmes* and *actions* (e.g. Programme Arachide; and in the OHVN zone, the Action Tabac) were being created within the Ministry of Agriculture (Diallo, 1990; Min. de la Production, n.d.). This first round of agricultural development, however, failed to achieve much impact due to a cumbersome national bureaucracy, lack of funds and shortage of trained personnel. By 1967, the government began to refine its approach through the creation of a number of separate, self-contained, semi-autonomous *opérations* that lay partly outside the control of central Ministries. Relying upon production packages similar to those developed by Viguier and used by the early *programmes* and *actions*, the new *opérations* were structured around specific agricultural systems for the production of key economic crops within well-defined geographic areas. Under this new approach, the future role of the Ministry of Agriculture was envisioned as being supervisory, presiding over a growing number of *opérations* and governmental agro-industrial *sociétés* (e.g. Société des Conserveries du Mali, SOCOMA) (Min. de la Production, n.d.).

One of the intentions behind the creation of these semi-autonomous development schemes was to attract external donors who could help with their financing (Min. de la Production, n.d.). In fact, the identification of external funding was one of three criteria used in creating new *opérations* (*ibid.*). The remaining two criteria, the creation of a trained cadre of Malian professionals to staff and manage the *opérations*, and the establishment of a budgeting system independent from that of the funding source, were designed to keep these *opérations* from becoming neo-colonial enterprises of exploitation, one of the principal concerns of the new government (*ibid.*).

In 1972, following the 1968 *coup d'état* and a period of economic recovery, the second post-Independent Malian government elevated its commitment to the *opération* approach to a new level, by making *Opérations de Développement Rural* (ODR) its official strategy of rural development. The ODRs were founded upon the same principles of administrative and financial autonomy, with the same mandate to pursue their development objectives 'through any means available' as found in earlier efforts. The planning documents of the second five-year Plan, 1974–78, clearly illustrate how central the ODRs were to meeting national development objectives (CNPER, 1972). Under the second Plan, the country was partitioned into 11 rural development zones, each of which roughly corresponded to the location of an ODR.[2] By the mid-1970s there were at least 18 agriculture ODRs and their smaller cousins, the *Actions*, functioning within the country (Diallo, 1990).

From an institutional standpoint, the creation of rural development zones, and the growing number of ODRs, marked an important turning point in the administration of agricultural and rural development in Mali. First, this initiative moved the majority of administrative responsibilities out from under the centralized, poorly financed and staffed, bureaucratic structures that had previously ground development efforts to a halt. Secondly, from a programmatic standpoint, the rural development zones and ODRs were established largely along agro-ecological lines. In determining their boundaries, the ODRs grouped together villages with similar agro-ecological conditions that could be easily serviced through a single development structure. The identification of areas with fairly homogeneous conditions allowed each of the ODRs to focus their efforts on exploiting the potential of a single set of conditions in the production of the particular commodity for which they were responsible (thus allowing the ODRs to focus on the use and perfection of one technical package).

In terms of attracting external funding, the ODR approach was highly successful. By the early 1970s, nearly 90 per cent of the financing for the different *opérations* came from foreign sources (e.g. European, French, among others) (Min. du Plan, n.d.). *Opérations* which did not have external funding tended to exist on paper only (Steedman *et al.*, 1976), as nearly all available government finances were used to meet the matching fund requirements of the donors. Through a pricing structure known as a *barème*, the ODRs were intended to become increasingly able to generate their own revenues. This system of subsidies and price controls, established through the governmental monopolies on input supply and marketing channels (e.g. Société de Crédit Agricole et d'Equipment Rural (SCAER); Société Malienne d'Import Export (SOMIEX); Office des Produits Agricoles du Mali (OPAM)), allowed the ODRs to capture commissions on input sales and through the marketing of agricultural commodities which they controlled.

The unique characteristics of each ODR also contributed to increasingly disparate rates of growth in the different rural development zones. Due to the combination of differences in the level of external funding, the profitability of the particular commodities being promoted, and the success of individual development approaches, each ODR has experienced its own trajectory in economic growth. Their independence from governmental support and dependence upon external financing have posed problems for the ODRs' longevity, as in cases where external support has been withdrawn (i.e. the demise of the OACV). By establishing a new set of organizational divisions—*zones*, *secteurs*, *secteur de base*—the ODRs introduced an administrative system independent from that established by the first Malian government.[3] This financial and administrative autonomy of the ODRs has led to conflicts between the ODRs and the comparatively poorly-funded local development efforts undertaken by the governmental ministries (Bingen, 1985).

APPENDIX C

The evolution of the OHVN extension programme

Phase I: The Opération Haute Vallée (OHV) Project

Following the decline in French and European funding in the mid-1970s, USAID began financing a new, large, highly ambitious, integrated rural development project in the Haute Vallée area in 1978 (Opération Haute Vallée (OHV)). The project's overall goal was to 'increase the income and improve the quality of life of the rural poor' through increased agricultural productivity and marketing (USAID, 1978). Designed around the major technical assumption that, aside from lack of irrigation, the limited use of animal traction (AT) was the major impediment to agricultural growth in the area, the project was composed of seven thematic areas:

- animal traction—promotion, training and research[1]
- modifications in the agricultural credit programme, to allow more farmers to invest in AT and other production inputs
- road construction and improvements to facilitate the marketing of increased agricultural produce
- a rural health-care programme to improve labour efficiency
- a functional literacy programme to improve farmers' ability to access and utilize new technologies
- rehabilitation and expansion of 640ha of the Bancoumana polder
- administrative and technical support of the OHV in providing services to area farmers (USAID, 1978).

In addition, minor components, such as a blacksmith training programme to increase local capacity in manufacturing and repairing AT equipment, were added. The combined impact of these components was expected to rapidly transform of agriculture in the OHV zone.

The Eighth Project Amendment

After stinging critiques of the project's early performance (e.g. USAID, 1982; RONCO, 1985), substantive steps were undertaken to reform and improve the project's management and financial operations. Originally funded under a five-year initial grant, the OHV project was extended and amended several times before it reached USAID's ten-year statute of limitation on project funding, and had received nearly US$ 20million in total financing. Any success, however, that the project experienced during Phase I is directly attributable to the conditions imposed by the Eighth Project Amendment. This agreement essentially redesigned the OHV project, reducing and refining the seven thematic areas to five: credit; functional literacy; road construction; on-farm research; and the administrative support of the OHV. The health and Bancoumana polder rehabilitation components were dropped completely, while the emphasis on animal traction research and training was replaced with a broader research focus on farming systems.

One major issue that neither appeared in the original project paper, nor in the Eighth Amendment, was the recognized need for or provision of technical assistance for improving the performance of the extension service. The early evaluations of the project (RONCO, 1985; USAID, 1982) were highly critical of the extension service's performance, calling it 'inefficient and ineffective' (USAID, 1982). Subsequent reviews (RONCO, 1985; USAID, 1986; Lebeau, 1986) continued to comment on the *opération's* weak extension service, but increasingly began to focus upon the closely related issue of the 'tired, old' extension themes and the lack of relevant, technical information. As Lebeau (1986) pointed out, the 'limits on available technology limits the impact of extension'. These problems were taken up in the project's redesign, under a second phase of funding.

Phase II: The development of the Haute Vallée (DHV) Project

Phase II of USAID's investment in the Haute Vallée area relied extensively upon the recommendations of the World Bank commissioned study on reorganizing the Malian ODRs (SATEC, 1985; cf. Min. du Plan, 1987). In designing the new Development of the Haute Vallée (DHV) project, the role of the OHV was redefined as primarily an extension and planning organization. The DHV project was to address three main areas: 'OHV restructuring, involving the gradual withdrawal from credit, input supply and marketing functions, accompanied by significant reductions in OHV staff and a strengthening of the OHV extension function; rural enterprise and rural institutional development, involving village-based farmer co-operatives, private business firms, and intermediate financial institutions; and expansion of the rural road network, to stimulate markets and private enterprises in the Haute Vallée' (USAID, 1988:6). Under this plan, USAID was also to continue financing the credit, rural roads and functional literacy efforts begun under Phase I.

In improving the extension (*vulgarisation*) programme, the primary measures included: a reconfiguration of the OHV administrative structure and significant cuts in field staff; redefined jobs and improved technical support for the remaining extension field agents; and an increased transfer of extension responsibilities to rural villages (see OHVN, 1988). The number of OHV administrative sectors (*secteurs*) increased from six to ten, and the four-tiered extension organization (headquarters, *secteurs, zone d'expansion rurale, secteur de base*) was replaced with a three-tiered structure (headquarters, *secteurs*, *sous-secteurs*). This reorganization was matched with a nearly 50 per cent cut in the number of extension field staff. At the same time, a new women's programme was created, along with the position of *animatrice* to co-ordinate the new programme's activities in each *secteur*. The OHV was also elevated in status by the Malian government to an *établissement publique administratif*, becoming the Office de la Haute Vallée de Niger (OHVN), and thus acquiring greater economic freedom.

During the latter half of the Phase I OHV project, the OHV had begun experimenting with a group approach for extension activities. Under Phase II, the OHVN increasingly applied, and eventually formally adopted a Training and Visit (T&V) system of extension (see Benor *et al.*, 1984). The major responsibility for agricultural credit, marketing and input supply was transferred from the purview of the extension service, and field staff jobs were redefined to focus solely on extension of technical information to farmers. A *division de vulgarisation* was created, as were seven new Subject Matter Specialist positions (SMS) to strengthen the training and technical backstopping of field-agents. The OHVN research programme, and the position of research co-ordinator, were phased out, as USAID began financing a separate farming-systems research project in the

OHVN zone (USAID, 1984) (see Appendix D). In addition to the support given to the creation of *associations villageoise* (AV), under the new governmental objective of increased 'responsabilisation du monde rural', the OHVN began to increasingly target the creation of self-managed extension villages (*villages auto-encadrés*) as a major goal in its programming. In the *villages auto-encadré*, the village *animateurs* (who are not paid by the OHVN) assume nearly all of the regular extension duties previously handled by the CSS, often co-ordinating their input and marketing activities with other nearby *villages auto-encadré*.

The first DHV project amendment

In 1993, the first DHV project amendment was issued, which extended the life of the OHVN project for an additional five years and increased total spending on Phase II from US$ 17.5million to US$ 29.5million (USAID, 1993). In addition to continuing support for the training of village associations, functional literacy training, credit scheme and rural roads construction, this amendment added a family-planning and AIDS-awareness component, and provided for an expanded emphasis on natural-resource management through the OHVN extension programme, including the addition of a pilot, community-based forest management initiative. An agribusiness and marketing unit was also added to the OHVN's operations. Coming full circle to the idea of fruit and vegetable exportation that was briefly promoted by the first independent regime in the early 1960s, the new OHVN agribusiness/marketing unit was charged with developing commodity export and domestic valued-added processing markets for farmers in the OHVN zone, as well as providing technical and managerial training for local business entrepreneurs.

APPENDIX D

The evolution of agricultural research in the OHVN zone

Shortly after Independence, the Institut d'Economie Rurale (IER) was created within the Ministry of Agriculture to help co-ordinate and direct the national agricultural research programmes. The rural development organizations (ODRs), however, continued to draw virtually all of their technical support directly from a number of French research programmes.[1] In addition, most of the research sections within the IER that dealt with issues of food-crop production were headed by expatriate French researchers on loan from one of the French research institutes (USAID, 1984). This pattern of reliance upon both the French research system and individual researchers continued until the late 1970s, when a growing cadre of trained Malian professionals, in collaboration with several regional and international research programmes, began to take a more active role in leading Mali's national research and development programmes. In 1979, the Division de Recherche sur les Systèmes de Production was formed within the IER to adapt research to farmers' conditions, and to help direct future research efforts in meeting national development goals (this division was later elevated to *Département* status, becoming the DRSPR). In the same year, the regional Semi-Arid Food Grain Research and Development project (SAFGRAD), financed largely by USAID, began conducting on-farm varietal trials in Mali in collaboration with IER scientists. The International Crop Research Institute for the Semi-Arid Tropics (ICRISAT), also supported by USAID funds, began actively collaborating with IER researchers in developing breeding and agronomy programmes (Schilling *et al.*, 1989).[2] The eventual separation of Malian research programmes from their French predecessors marked an important turning point in Malian agricultural research, and fuelled the growing capacity of the country to design and carry out its own research agenda. By the beginning of the 1990s the IER boasted one of the largest national research systems in West Africa, with over 130 trained researchers (ISNAR, 1990).

Prior to the changes occurring in the national research structure during the late 1970s, a few ODRs, including the Opération de la Haute Vallée (OHV), which were not closely aligned with any of the commodity programmes receiving direct research support from the French, began to develop their own independent research units. When USAID began financing Phase I of the OHV project in 1978 (see Appendix C), a research component was created which focused on issues related to advancing AT technologies. By 'using the assistance of Peace Corps Volunteers and two American agronomists... to test ploughs, plant spacing, varieties, planting dates, weeding techniques, etc.' the research programme gradually evolved into a larger on-farm testing and demonstration effort (RONCO, 1985). These research efforts were considered 'a critical element of the project because it [was] the only component which [would] eventually enable the... programme to move beyond its initial focus on animal traction' (USAID, 1978).

In 1983, the Eighth Project Amendment began to shift emphasis within the OHV research efforts towards a broader farming systems research and extension

(FSR/E) focus. Separate contracts were negotiated with the DRSPR and SAFGRAD programmes respectively to undertake socio-ecological studies and conduct varietal trials within the Project zone that would help the OHV to target its extension activities more effectively, and deliver higher-producing, new varieties to farmers. The position of a research co-ordinator was added to the OHV administration to help co-ordinate the OHV's own research efforts with those of the DRSPR and SAFGRAD, as well as the growing number of other Malian research divisions that were beginning to conduct research in the zone (e.g. Section de Recherche sur les Cultures Vivrière et Oléagineuses, SRCVO; Section de Recherche Coton, Fibre et Jutières, SRCFJ; Cellule Essais Multilocaux, CEM) (Kagbo, 1986).

The SATEC study on the restructuring of the ODRs, presented in 1985, had a major impact on the future of agricultural research within the OHVN (SATEC, 1985). Among its many recommendations, the study strongly argued that ODRs, such as the OHVN, which managed their own research programmes, should terminate these separate efforts in order to reduce needless repetition and to concentrate research expenditures at the national level where all of the country's development programmes could benefit (SATEC, 1985). In 1985, USAID funded the creation of a new section (*volet*) within the DRSPR, the DRSPR *Volet* OHV (DRSPR/OHV), to work exclusively on agricultural improvements in the OHVN zone. This heralded a change from earlier attempts to increase the research capacity within the OHVN organization itself. In 1988, with the début of the Phase II DHV project (see Appendix C), the OHVN officially closed its research section and established the position of Subject Matter Specialists (SMS) to serve as liaisons between the OHVN and the DRSPR/OHV and other governmental research organizations. Thus ended the period of the OHVN's direct involvement with on-farm research and testing. From this point forward the OHVN became entirely dependent upon other agencies and institutions for the development of new technical information and extension messages. Chief among these sources of information was the new DRSPR/OHV farming-systems research programme.

The DRSPR is now known as the Programme Système de Production et Gestion des Ressources Naturelles (PSPGRN). This change in the DRSPR, and other initiatives, are part of the USAID-financed Strengthening Research Planning and Research on Commodities Project (SPARC), and related efforts (e.g. ISNAR's long-term support of IER's management capacity; the multi-donor Special Programme for African Agricultural Research (SPAAR); and the IDA-funded National Agricultural Research Project (NARP)). These initiatives are broadly aimed at improving agricultural research in Mali, and redressing the debilitating trend of individually funded projects dominating the national research-system agenda.[3] Two of the major aims of the SPARC programme have been to decentralize the research system into six regional research centres (up to 90 per cent of the professionally trained research staff had previously resided in Bamako), and eliminate the various Departments and restructure the IER around a number of research programmes (an even larger number of sub-programmes) implemented through the regional centres (USAID, 1992).

NOTES

Chapter 1

1 The rate of growth in volume of cotton produced fell from 16.5 (1973–80) to 2.2 per cent (1980–87); growth in groundnut oil production fell from 22.7 (1973–80) to −25.7 per cent (1980–87) (World Bank, 1989).
2 Although these general trends in Mali appear to support the larger view of a growing African agricultural crisis, a closer look at national indicators, particularly with reference to the agricultural sector, reveals a number of contradictions. On the one hand, a comparison of the five-year periods 1985–89 and 1990–94 shows that although the land area under total crop production increased by 20 per cent, average cereal yields fell by 22 per cent (World Bank, 1997), while from a longer-term perspective, the 25-year period between 1970 and 1995 shows that while food production capacity (5 yr yield average × area under production) increased by 53 per cent, the country's population over this period doubled (World Bank, 1997). However, if these figures are to be believed—and there is mounting evidence to question their veracity (see Wiggins, 1995)—one would expect to see steeply rising levels of food imports and food aid, coupled with possible mass migration and increasingly widespread incidence of malnutrition. Yet there is little evidence supporting such a trend. In fact, having peaked in the mid-1980s, the combined volume of cereal imports and food aid have dropped to well below their 1978 levels, declining 40 per cent alone between 1985–89 and 1990–94 (World Bank, 1997). Rates of infant mortality have continued to drop since 1980, and per capita daily calorie intake reportedly increased during the height of the crisis (1980–86) (World Bank, 1989; 1995).
3 Nevertheless, in 1991 research expenditures in Mali barely exceeded one per cent of the agricultural Gross Domestic Product (GDP), with nearly two-thirds of this investment being supplied by donors (Pardey *et al.* 1997).
4 Cereal crop yields for the three-year periods of 1970–72 and 1993–95 (the oldest and most recent data available in World Bank African Development Indicators) (World Bank, 1997) averaged 782kg/ha and 792kg/ha, respectively.
5 Roughly 500,000 people live in the OHVN. Data from the 1987 National Census placed the rate of rural population growth within Mali's Second Region (within which the OHVN is located) at 2.28 per cent (Davies, 1996), the highest in the country. Many of the immigrants to this region have come from the northern parts of the country, where, due to the reduction in rainfall and insecurity associated with the Tuareg insurrection, population levels reportedly declined by 5 per cent between 1980 and 1992 (République du Mali, 1994).
6 For example, between 1980 and 1990, the number of hectares in the OHVN planted to improved maize varieties grew by 66 per cent, while in neighbouring and agro-ecologically similar areas, the increase was roughly four times as high (Boughton and Henri de Frahan, 1992; Kingsbury *et al.*, 1994).
7 This figure has been adjusted to eliminate the influence of the disastrous 1990 harvest; the 1991 harvest surpassed that of 1990 by 140 per cent.
8 See, for example, the following bibliographies and collected works: Agriculture and Human Values, 1989 and 1991; Brokensha *et al.*, 1980; de Boef *et al.*, 1993; Gupta, 1990; IDS, 1979; Inglis, 1993; Johannes, 1989; McCall, 1995; Mundy, 1991; Narby and Davis, 1983; Warren, 1991; Warren *et al.*, 1989 and 1995; cf. back issues of the *Indigenous Knowledge and Development Monitor*; publications of the Technology and Social Change programme at Iowa State University; other volumes in the Intermediate Technology Publications Series on Indigenous Knowledge; and the searchable, on-line CIKARD Citation Index: http:/www.physics.iastate.edu/CIKARD/database.html).
9 The research upon which this book is based was conducted as part of a contractual agreement between the Department of Resource Development at Michigan State University and the United States Agency for International Development (USAID).

The scope of the study included a detailed technology assessment of the importance and impact of both the indigenous and introduced practices, crops and processes of innovation and communication on changes in the agricultural systems of the area; an assessment of the future technical service needs of farmers and village associations in the area; the identification and assessment of the contributions to rural development by other actors; the organizational dynamics of the village associations and their ability to act as 'engines' for local development; an assessment of the changing political context within which development efforts were taking place, and implications for future activities; and the identification of organizational constraints on further development efforts (see Bingen *et al.*, 1994).

10 Contrary to Drinkwater's interpretation (1992), the Grounded Theory approach neither explicitly assumes complete objectivity on the part of the researcher, nor takes a positivistic view of reality (see Strauss and Corbin, 1990).

Chapter 2

1 First used to describe a specific vegetative zone south of the Saharan desert (Chevalier, 1900), the Sahel has become associated with a larger regional image of a harsh, transitional environment (Doenges, 1988; Franke and Chasin, 1980). Sahel is an Arabic word meaning literally 'shore' or 'border,' and is used in reference to the Sahara, which is derived from the Arabic *sahra*, meaning 'wilderness.' The Sudan, the area south of the Sahel, is derived from the Arabic word for 'black' and its use originates in reference to the people inhabiting this region (from Franke and Chasin, 1980).

2 There is no universally accepted definition for these climatological and vegetative zones (e.g. Davy *et al.*, 1976). The limits adopted here are in general agreement with the original definitions proposed by Chevalier (1900), as described by Lawson (1986).

3 The following seven countries are generally included in reference to the Sahelian West Africa: Senegal, Gambia, Mauritania, Mali, Burkina Faso, Niger and Chad (MacDonald, 1986), to which can be added the northern portions of Nigeria.

4 As the ITCZ moves northward, the moist ground flow is contained beneath the region's dominant dry trade winds coming out of the north-east. Regular disturbances in these overlying trade winds allow convective activity in the moist surface-flow to break through the containing dry air mass, forming cloud clusters which are responsible for the region's rainfall. Once formed, these storm events flow over the region from north-east to south-west, pushed by the north-easterly trades (Farmer and Wigley, 1985).

5 The main rainfall zone generally follows 450–500km, or four to six weeks, behind the ITCZ (Cochemé and Franquin, 1967; Farmer and Wigley, 1985).

6 The coefficient of variation—defined as the standard deviation of annual totals expressed as a percentage of the annual mean—describes the variation about the mean that is expected two-thirds of the time (Nicholson, 1982). For example, a 50 per cent variability means that yearly rainfall will fall within 50 per cent of the annual mean two-thirds of the time.

7 The cloud clusters that form along the ITCZ can be over 1000km wide, with line squalls of actual rainfall averaging 5–20km in width (Davy *et al.*, 1976; Farmer and Wigley, 1985).

8 For example, the passage of one storm in Niger was observed to shift 100km during the first six-hour period, 400km during the second six hours, slowed again, then travelled 600km in the final six hours before dissipating entirely (Hayward and Oguntoyinbo, 1987).

9 In any one year, 50 per cent of the rainfall stations in the Sahel report rainfall below long-term averages (Nicholson, 1982)

10 For the major soil groups found in Mali, infiltration rates average 10mm/hr, while precipitation levels exceed 27mm/hr in nearly 50 per cent of the storm events (Sivakumar, 1989).

11 For an examination and comparison of the extensive social impacts of this period and the severe 1910 drought, see Kates (1981).

12 Nicholson (1982) notes that, as a rule of thumb, in the Sahel a 100mm shift in rainfall levels equates to a 100km movement along the North–South axis. The positioning of the rainfall isohyets is extremely sensitive to the measurement period being used. Although the World Meteorological Organization (WMO) has adopted the period 1960–91 as its official measurement standard, many maps in use are based upon the earlier WMO standard that covers the period 1930–61, or contain only a portion of the present drier period.

13 A two-week delay in planting can result in yield losses of up to 30 per cent for peanuts and 60 per cent for sorghum (ICRISAT, 1978), and 50 per cent for cowpeas (Ntare and Williams, 1992).

14 Similar observations of extensive, drought-induced mortalities in *A. albida* have been noted in neighbouring Senegal (Seyler, 1993). In some areas, it is estimated that 40 per cent of this species perished during the 1968–73 drought (Wentling, 1983 in Seyler, 1993), while other species suffered losses of up to 60 per cent of their populations (Bradley, 1977).

15 Including the southern administrative *secteurs* of Bancoumana, Kangaba, Ouelésse-bougou, Dangassa and Gouani (PIRT, 1989a). Soils in this area are broadly classified as alfisols and ultisols under the American system (Steila and Ponds, 1989), or as ferruginous, ferrisols and ferrallitic soils in the UN classification scheme, and as 'sols ferrugineux tropicaux non ou peu lessives' in the French system (Ahn, 1969; Charreau, 1974).

16 Inselbergs are isolated mountainous outcroppings (which in the OHVN are typically composed of schist), that have become deeply buried in their own debris. Soils developing out of these scattered schist deposits have a higher nutrient status than the surrounding soils. In general, the differences in parent material and intensity of weathering strongly influence the formation of different clay types, with important implications for soil fertility in terms of their ability to retain nutrients in the soil profile (i.e. the cation exchange capacity (CEC)). Variable montmoullorite-type clay developing out of schist deposits has a CEC of 100–130me/100mg, roughly 50 times that of the more strongly weathered kaolinitic clays associated with the granitic 'basement complex' formations, which have a CEC of 2–6me/100mg (Ahn, 1969).

17 Including the administrative *secteurs* of northern Bancoumana, Koulikoro, Sirakorola, Banamba and Boron (PIRT, 1989b). The majority of soils in this northern area are classified as alfisols under the American system (Steila and Pond, 1989), corresponding to ferruginous soils and brown and red soils of arid and semi-arid areas in the UN system, and 'sols rouge et brun' in the French system.

18 Although these well-drained, predominantly sandy soils generally contain little organic matter, they have a larger clay fraction near the surface, comprised mainly of montmoullorite clays, giving them a higher CEC and therefore greater potential of retaining the available nutrients than those soils found further to the south (Jones and Wild, 1975).

19 In the Sahelian and Sudanian environments of West Africa, as much as 70 per cent of the above-ground woody biomass is thought to pass through termites. The short and long-term management implications of this are poorly understood and possibly underappreciated (Josens, 1983). The density of termite mounds varies by area and species. In Nigeria, mound densities average 32/ha, equating to roughly 2–3 per cent of the land surface. Termites can bring to the surface significant amounts of clay that have been translocated from the surface soil horizons. In addition to contributing to moisture retention, these clays can also increase the CEC of the soil, improving the nutrient-holding capacity of the areas immediately adjacent to the termite mounds (Miedema and van Vuure, 1977).

20 Studies completed in Senegal during the 1968–73 drought showed that while rainfall in the area had dropped by 45 per cent, herbaceous biomass production initially fell only 20 per cent, although as the drought continued, biomass production dropped to virtually nil (MacDonald, 1986).

21 The higher population density of this area is associated with the historical prevalence of 'river-blindness' (onchocerciasis) and livestock diseases in the moister regions of the south.

Chapter 3

1 Under conditions of severe drought, farmers may be forced to abandon much of their regular crop production activities and devote their remaining energies and resources to collecting wild grains, fruits and vegetables. Most of the older farmers interviewed told of having had to rely upon 'wild foods' (e.g. *Paspalum spp.*, *Dioscorea spp.* amongst others), and the now familiar anecdote of raiding ant and termite stores to secure grain, during the worst of the droughts. They said that this was of declining necessity, however, because 'the trucks' now arrive to deliver relief food during these times of crisis.

2 In another illustration, the highly integrated agroforestry gardens found in the southern parts of the zone are often manured with kitchen wastes and planted to thin stands of cereals. The household wastes contain a large number of viable seeds (a fact known by farmers) from a number of different crops (the composition of which is not known), that are then allowed to prosper wherever they sprout. The result is a randomly 'intercropped' plot which may contain over a dozen different varieties, giving the appearance of a highly integrated cropping system, yet produced almost entirely through chance.

3 The total population of the OHVN zone, roughly 500,000, is divided among three dominant agricultural groups and one pastoral group. The Bambara are by far the most numerous, occupying the central part of the zone, with the Malinké concentrated in the southern and western-most *secteurs* of Bancoumana, Kangaba and Kati (BECIS, 1991), and the Sankoré populating much of the northern *secteurs*, Boron, Banamba, and parts of Sirakorola (BECIS, 1991). The pastoral Peuls are found residing throughout the zone. Several authors note that there is little difference in the social structures of the Bambara and Malinké, and have accordingly treated them as essentially the same (Koenig, 1986a; McConnell, 1993; Moseley, 1993).

4 There may be some significant differences in the social organization of Bambara and Malinké households in terms of the division of land and labour with respect to gender. Grigsby (1989) reports that, in general, women in Malinké households have larger personal fields, more secure control and greater freedom to invest labour in these fields. In contrast, women in Bambara communities may compensate for their lack of security over private fields by managing joint fields with other household women, or by working in the village women's group, through which they are able to leverage more secure land and labour rights.

5 Karité fruit production fluctuates widely between years. Although the cause and control of this variance has as yet eluded researchers and farmers alike, it has been noted that late dry-season fires can damage the flower set of the trees, affecting their productivity (e.g. SPORE, 1991; cf. Boffa *et al.*, 1996).

6 Open access resources are those for which there are no functional rules of access or use, while common property refers to those jointly owned resources where a group exercises exclusionary control and, in the case of common property resource management, may enforce rules of conduct (e.g. NRC, 1986).

7 The northernmost portion of Boron *secteur*, with its extremely low rainfall, may present an entirely different set of opportunities that this research does not cover.

8 The significantly different agro-ecological conditions of the Mandingue plateau area (mainly Kati *secteur*) may be the basis for identifying a separate portfolio.

9 Throughout the OHVN zone, villages located on both banks of the Niger River have distinct production opportunities which may necessitate the description of a separate 'River Portfolio', although insufficient data prevent the presentation of an accurate description of its characteristics here.

10 Vegetable production was promoted in this area during the mid-1960s, as was special support for tobacco production, which continued up through the 1970s. However, the decline in rainfall and producer prices, changes in institutional support and the emergence of other opportunities have led to stagnation and decay in the tobacco subsector.

Chapter 4

1 For example, Allan, 1965; Brokensha *et al.*, 1980; de Schlippe, 1956; Francis and Sanders, 1978; Gliessman *et al.*, 1981; Kater *et al.*, 1992; Kessler, 1992; Rao and Wiley, 1980; Richards, 1985; Thurston, 1996; Warner, 1991; Warren *et al.*, 1989; 1995; amongst many others.
2 e.g., Altieri, 1987; 1990; Altieri and Hecht, 1990; Carroll *et al.*, 1990; Francis, 1986; Gliessman 1990; Innis, 1997; Vandermeer, 1989; amongst many others.
3 See, Aubert and Newsky, 1949; Bradley, 1983; Diallo, 1991; ICRISAT, 1989; McConnell, 1993; Moseley, 1993; Niamir, 1990; Pawluk *et al.*, 1992; Richards, 1979; Sikana, 1993; Warren, 1992; cf. Inglis, 1993; Johannes, 1989; amongst many others.
4 Using international systems of classification, studies on the biological diversity of Malian fauna have identified 950 species of mammals, fish (in the Niger basin), reptiles (data on amphibians is lacking) and birds (Warshall, 1989; WRI, 1994), a figure remarkably close to that of the traditional assessments of different life-forms.
5 Such a system of classification is not without parallel in Western science. The US Dept. of Agriculture's Soil Texture Triangle, for example, is another texture-based classification scheme that is organized in relational terms. Field use of this classification scheme requires significant knowledge of a range of soil types and the mastery of tactile 'hand' tests to help categorize soils. The categories themselves, however, unlike those in indigenous taxonomies, are ultimately defined by specific percentages of silt, clay and loam, which do not alter by location.
6 Not surprisingly, the principal cereal crops indigenous to West Africa—millet, sorghum, fonio and to a lesser extent African rice—are generally tolerant of acidic soils (millet can tolerate pH as low as 4 to 6, while sorghum can withstand pH of 5 to 8), and in the case of millet (and fonio), can stand high levels of soil aluminum (NRC, 1996).
7 In another example, it can safely be said that farmers throughout the world are often unaware of the magnitude of their erosion problems. In a study conducted in the US, less than 2 per cent of farm managers, and none of the land owners, identified severe erosion problems, whereas the Soil Conservation Service identified 82 per cent of the farms as having major erosion problems (Batie, 1983).
8 Through direct facilitation and a more complete exploitation of the available resources, intercropping systems are said to 'over-yield,' in reference to their ability to produce more on a per-unit basis than if the same crops are grown in separate monoculture stands.
9 The competitive production principle refers to the more complete exploitation of available resources through inter-species competition, while the facilitative production principle refers to the actual transfer of benefits from one species to another.
10 As Lewis (1979) notes in differentiating between offspring and descendants within Bambara society, it is the successful integration of offspring into the network of social relations and obligations that gives rise to true descendants. This process is reflected in part by individuals' reference to themselves as 'Bambara' rather than 'farmers' (Lewis, 1979), with the implication that being Bambara is more than just a birthright, but is an achieved status (Grosz-Ngaté, 1986).
11 See Lewis (1979) for a discussion of the established 'marriage-routes' within his study area. The inter-family transfer of information and agricultural practices along marriage lines is also noted by Knight (1974) in his study of agricultural change in Tanzania.
12 Although not covered in this research, the Peuls and other pastoral groups possess a tremendous amount of specialized knowledge and expertise in herding animals (e.g.

Mathias-Mundy and McCorkle, 1989; McCorkle *et al.*, 1996; Niamir, 1990; Slaybaugh-Mitchell, 1995).

13 Lewis (1979) argues that counter to Zahan's (1960; 1974) contention, the different initiation groups may not represent a progressive continuum, but may, in fact, constitute independent entities.

14 Millar (1996) also notes the importance of competitive aspects of agricultural knowledge accumulation and performance in his study from Ghana.

15 Lewis (1979) refers to the 'joking' relationships that commonly develop between men and women, which provide a socially sanctioned medium through which individuals of different gender can talk with one another. Friendships between members of the same sex can also evolve into highly codified relationships, e.g. *amis intimes*, or blood brother-type relationships (pers. com. McCorkle).

16 In local markets, individual vendors offered small quantities of seed stock for sale separately from grain meant for consumption. The purchase of seed for planting from larger merchants is considered by farmers to be less desirable than home-grown seed, because of the poorer quality, lack of genetic purity and often unknown provenance.

17 Findings from the TROPSOIL Collaborative Research Support Programme (Anon., 1992) show that residue retention, in addition to providing the well-known benefits of reducing surface temperature and wind speed, helping to curb water erosion and increasing moisture retention, and adding needed biomass to the soil, also serves as such an efficient trap of ground flow and *harmattan* dusts that it actually helps to increase soil depth. (This wind-blown dust is a significant, and possibly underappreciated, source of soil nutrients, e.g. Jones, 1938; McTainsh and Walker, 1982; Scott-Wendt *et al.*, 1988).

Chapter 5

1 Along with the rest of the Malian research system, the DRSPR has been restructured and is now known as the Programme Système de Production et Gestion des Ressources Naturelles (PSPGRN). However, due to the historical nature of this analysis, and to avoid confusion, reference will continue to be made to the DRSPR/OHV throughout.

2 By the mid-1970s, pressure from the international donor community to adopt integrated approaches to rural development (e.g. UN, 1971) led other Malian ODRs also to begin diversifying and expanding the range of services that they provided to farmers (see World Bank, 1974).

3 To avoid confusion, the names of the USAID-funded projects—the Phase I Opération Haute Vallée (OHV) and Phase II Development of the Haute Vallée (DHV)—are not used (see Appendix B). These project titles are distinct and separate from the titles of the Malian ODR, the Opération de la Haute Vallée (OHV), and its subsequent elevation in status to the Office de la Haute Vallée du Niger (OHVN). While the former acronyms denote USAID-funded projects, which include not only the financing of the ODR but a range of additional activities, the latter acronyms denote the ODR's economic and administrative relationship to the national government.

4 The International Labor Organization (ILO) had assembled many of these technical packages, based largely upon technologies developed under the colonial system, into manuals that served as the primary references of the ODR technical programmes for many years (Steedman *et al.*, 1976) (compare CIRAD, 1980; IER 1990; IRAT, 1975).

5 There is generally believed to be over 60 national, international, bilateral and non-governmental funded projects and programmes operating in the OHVN zone. Taken as a whole, the various governmental programmes, and bilateral and international projects offer farmers a range of technical services, but individually tend to be very limited in scope and geographic range of coverage. The vast majority operate in complete isolation from the OHVN extension service. Neither the government's Cellue de Suivi des ONGs, nor the donor-financed Comité de Coordination des Actions des ONG (CCA/ONG), maintain complete or accurate records on the plethora of NGO activities

found in the Zone. Although at one time the OHVN operated a NGO co-ordinating office, this has long since been closed. Presently, the NGOs operating in the OHVN zone tend to deal with quality-of-life concerns, focusing on improvements to village infrastructure (e.g. wells, health units, schoolrooms, small-scale irrigation structures), and a diverse range of small-scale economic activities (e.g. gardening, bee-keeping, animal husbandry, cereal mills and *karité* presses), many of which are targeted at women. Few, if any, conduct their own research and when confronted with technical problems, tend to consult governmental and other project technical units.

6 Under this project, USAID supported the construction of the national DRSPR headquarters at the Sotuba research station on the outskirts of Bamako, as well as the creation of two FSR field units. The first field unit, the DRSPR/OHV, became operational in 1986, followed in 1991 by the creation of the DRSPR/Mopti (USAID, 1984).

7 Knowingly or not, this typology is virtually identical to that developed by the OACV in the 1970s (e.g. IER, 1977).

8 Problems with the OHVN's technical programme have been noted throughout its history, e.g. Bingen, *et al.*, 1994; Kagbo, 1986; Lebeau, 1986; OHVN, 1988; 1989; 1991a; 1991b; RONCO, 1985; Sélingué, 1992; Simpson, 1995; USAID, 1982; 1986.

9 Through the mid-1970s, lightweight ploughs suitable for use with donkeys had been imported and sold to farmers in the Mopti area by the Catholic mission. Virtually all other ploughs made and sold in the country are of heavier construction, intended primarily for use with bullocks.

10 This is despite the explicit recognition in early project documents of the need to avoid stifling farmer creativity (USAID, 1978)

11 Fewer than a dozen *animatrices* (female extension agents) are responsible for working with women across the entire OHVN zone, compared with the 80 CSS who work exclusively with male farmers (the CSS reported spending less than one per cent of their time working with women prior to the creation of the women's programme (OHVN, 1990).

12 Earlier research had been conducted on the use of tied-ridges as a means of conserving soil moisture. However, this was abandoned after farmers raised the common complaint of the high labour demands associated with the technology. The proposed research of developing mechanical alternatives for the 'tie-ing' process appears never to have been carried out.

13 Recent research findings indicate that an even more dynamic appreciation of the soil environment may be required in order to optimize productivity in environments with highly variable rainfall (e.g. Brouwer *et al.*, 1993). This research has shown that some soil topographic features exhibit an inverse relationship in terms of their crop-yield response to rainfall, i.e. the least fertile soils may support the highest yields under poor rainfall conditions. The use of fertilizer on cereal crops, when indiscriminately applied, has repeatedly been shown to be uneconomical in environments with frequent shortfalls in precipitation, whereas the judicious application of fertilizer on crops in certain soils, even in very poor years, may in fact be economically viable.

14 The 16 *fiches* issued thus far include five on the various uses of rocklines and other structures, and six on tree planting (DRSPR/OHV, 1993b).

Chapter 6

1 Problems associated with household cash flow and the 'learning curve' of efficient managerial operation affect not only the rates of adoption of AT systems, but also the level of proficiency achieved by farmers using them. It is estimated that farmers require nearly five years of experience with an AT system before they are able to reap the full benefits of the technology (Jaeger, 1986). For farmers on the financial margin, five years of continued ownership of a functional AT team may be difficult to achieve.

2 A simple change in draught power sources, i.e. from manual to animal traction, is not accompanied by a yield response (Pingali *et al.*, 1987), although changes in cultural

practices, or timeliness, facilitated or associated with the use of AT can significantly affect yields (e.g. Charreau, 1974).

3 Village blacksmiths typically create templates of the common parts of each implement's design and use these for making replacement pieces and entire units from scratch. Thus, while the quality of the base materials may change due to local availability, the integrity of the basic design is preserved. Blacksmiths and farmers, however, were not extensively interviewed concerning their modifications of equipment design, thus the situation may not be as static as it appears (see, for example, Mills and Gilbert, 1990).

4 In only one village did farmers report the widespread use of anything but the traditional mound culture during the preceding generation.

5 Unlike farmers in other areas, those in the OHVN did not report any major adaptations or modifications of their hand tools (e.g. Coughenour and Nazat, 1985; McCorkle *et al.*, 1988; Nazat and Coughenour, 1987).

6 In areas outside of the OHVN, where mound culture is still widely practised, the innovation of 'tied mounds,' analogous to tied ridges, has been developed by DRSPR/Mopti, and is gaining widespread acceptance among area farmers (Kingsbury *et al.*, 1994).

7 Some of farmers' preferences may be explained by the method of application of the different fertilizers. Organic fertilizer is typically broadcast over the field before land preparation, whereas purchased inorganic fertilizers are generally applied to individual plants/rows after emergence (once it is obvious that the plants will survive), by which time striga has already established itself (see Riches *et al.*, 1993).

8 In addition to economic considerations, farmers in these northern areas also observed that fertilizer use tends to make their crops more susceptible to drought. It has been suggested that this phenomenon may be related to the early vegetative growth promoted by fertilizer use in phosphorous-deficient environments, which increases a plant's moisture demands, making it more vulnerable to drought damage caused by the common irregularity of early-season rainfall.

9 Farmers in the South-east Portfolio area reportedly stumble across old stonelines as they bring fields out of long fallow (Moseley, 1993).

10 Manure sales are often associated with economic need, or the sheer inability of some farmers physically to manage the large volumes of material, which commonly measures in the tons and requires significant labour, as well as ownership of or access to a cart for transport.

11 The timing of planting is particularly critical. A two-week delay in planting can result in a lowering of yields in peanuts by 30 per cent (ICRISAT, 1978), and as much as 50 per cent for cowpeas (Ntare and Williams, 1992) and 60 per cent for sorghum (Jaeger, 1986).

12 This programme provides planting recommendations to farmers in relation to locally monitored rainfall levels. Recommendations are based upon 30 years of rainfall and yield data, and are said to provide farmers with an 80 per cent probability of obtaining a harvest.

13 On-farm trials show that significant yield increases are possible when optimizing the timing of field operations (planting, thinning, weeding) (OHVN, 1993a). These demonstrations, however, have been confined to small-scale (0.5 ha) plots in isolation from the whole farm system where, because of labour and other resource constraints, farmers are unable simultaneously to optimize the timing of all the operations in each of their fields.

14 With this in mind, the formal research programme could possibly draw some important conceptual lessons from the development of 'precision' farming techniques, which through the use of massive field-level data sets and computer-aided technology, have attempted to bring to industrial agriculture the same sort of cost-effective and yield-improving practices that small-scale subsistence-level farmers have been able to achieve through their intimate knowledge of their production systems. While the technologies employed are clearly inappropriate to Malian farmers, the underlying concept of supporting tailor-made efficiencies is not.

15 Yet even these faster-maturing varieties are not impervious to damage wrought by the severest droughts. Farmers in some of the northern-most areas of the zone reported that during the latest period of drought, they were forced to consume their remaining seed stock (that which was not destroyed in the field after several re-plantings) in order to survive. This included many of their shortest season varieties, and in at least one location an entire species (e.g. fonio) which, two years afterwards, farmers had still not been able to replace (cf. Sperling and Loevinsohn, 1993). In this case, farmers eventually 'sent' a representative to a regional marketing centre to purchase a sack of fonio.

16 The improvement of local varieties has been a major theme within the ICRISAT and IER breeding programmes since the late 1970s. The most broadly adopted varieties of maize and sorghum emerging from the formal breeding efforts have been 'improved' local varieties (e.g. the sorghum varieties CSM-388 and *Tiémarifing*, and maize varieties *Tiémantié* and *Zanguéreni*).

17 Researchers at ICRISAT have been aware of a number of local, striga-resistant varieties since the late 1970s (ICRISAT, 1979). In trials, *seggatana*, or *seguetana*, has shown itself to be nearly three times as resistant to striga as the most effective 'improved' variety that researchers have developed (and is over 30 times more resistant than non-resistant 'improved' varieties) (ICRISAT, 1985).

18 Several improved varieties of millet offered to farmers through the research programmes and OHVN extension service have been widely rejected because of their susceptibility to bird damage (e.g. Schilling *et al.*, 1989). Traditional methods of bird control—chasing and the use of 'scarecrows' (pieces of metal and cloth strips tied to poles)—are labour intensive and quickly lose their effectiveness as the birds learn that they have little to fear.

19 *Pourghère* is used as a major ingredient in local soap manufacture and traditional medicines, and oil extracted from its seeds can be used as a substitute fuel in diesel engines, among other uses (e.g. Henning, 1989; 1992).

20 In the DRSPR/OHV research villages, the overall use of inorganic fertilizers on cereals is reported to be as much as 30 per cent of the total area planted. However, such findings cannot be generalized to the OHVN zone as a whole; much of the fertilizer applied in these research villages is brought to farmers' 'doorsteps' through a credit and delivery system managed by the extension service as part of the on-farm trials.

21 Research from Kenya underlines the extent to which agrochemicals pose a health hazard to small-scale farmers (Mwanthi and Kimani, 1993).

Chapter 7

1 Another way of approaching individual differences is through the examination of the 'styles' of farming that individuals from relatively similar opportunity sets choose to follow (see van der Ploeg, 1994).

2 When applied in a developmental sense, the process of 'naming the world', advocated by Freire (1970) and others (e.g. Fals-Borda and Rahman, 1991) is an empowering one, directed at breaking the 'culture of silence' by helping adults to move non-communicated and implicit knowledge into a conscious, communicable and action-enabling form.

3 De Schlippe (1956), among others, has discussed the one-ness of agriculture and culture in subsistence agrarian societies.

4 As noted by Mundy and Compton (1995), knowledge and information are not synonymous with one another. Knowledge is a socially influenced, personal construct, and 'it cannot be communicated [in its entirety] but is created in the minds of individuals as a result of the person's perceptions of the environment or through communication with others' (Mundy and Compton, 1995:112; cf. Dissanayake, 1986).

5 Niamir (1990:16), in her description of traditional pastoral management systems, describes these sources as '(1) accumulated cultural knowledge (2) knowledge modified through contact with other cultures, and (3) progressive learning of the environment.'

Other authors have identified similar or closely related relationships and sources, e.g. Barnett, 1953; Bell, 1979; Knight, 1974; McCorkle *et al.*, 1988; and more recently, Simpson, 1994; Blaikie *et al.* 1996; Campilan and Prain, 1997.

6 For a broader discussion on different forms of 'vertical' and 'horizontal' learning, see Millar (1996).

Chapter 8

1 For a discussion of recommendations targeted specifically at the public sector research and extension programmes operating within the OHVN see: Bingen *et al.*, 1994; Bingen and Simpson, 1997; Kingsbury *et al.*, 1994; McCorkle *et al.*, 1993; Simpson, 1995.

2 In doing so, R&E programmes must remain mindful of the need for strategic research in anticipating emerging problems and developing new opportunities (Maxwell, 1986; cf. Biggs and Clay, 1981).

3 See for instance, AZANIA, 1989; Innis, 1997; NRC, 1991, 1992, 1993, 1996, forthcoming; Reij *et al.*, 1996; Warren, 1991; Warren *et al.*, 1995; amongst many others.

4 This is contrary to the conclusion reached by Sumberg and Okali (1997) of the low potential for synergy in farmer–research interactions, due to the similarity of research processes used by each. Even in cases where farmers and researchers are examining similar technologies in the same manner, they are often doing so for very different reasons, motivated by different rationales and objectives, and bringing to bear different knowledge and resources. Innovation is as much about ideas as tools and procedures, and as such can be facilitated only by increasing the circulation of different perspectives and insights.

5 For example, in the Gambia (Mills and Gilbert, 1990; Willis *et al.*, 1995); Mali (Gubbels, 1997); Rwanda (Loevinsohn *et al.*, 1994; Sperling, 1994); Zaire (Mulume, 1997); Zambia (Drinkwater, 1994; Sikana, 1994); Botswana (Heinrich, 1993; Norman *et al.*, 1988; cf. Arnaiz, 1995; Eponou, 1996).

6 Although Lewin's original concern, with respect to training, was to develop a cadre of professionals capable of engaging in action research, the history of applying action research principles among community development professionals shows that developing the skills and organizational capacities of local groups is of equal, if not ultimately greater, importance.

Appendix A

1 I differentiate between RRA and Participatory Rural Appraisal (PRA) based upon the underlying objectives of each approach; RRA is concerned primarily with facilitating the learning of 'outsiders,' while PRA is more oriented towards 'insider' capacity-building as part of a larger change process (cf. Chambers, 1994a). Despite the general confusion over the use of the terms RRA and PRA, and the pejorative view that has come to be associated with RRA in the 1990s (i.e. the often false notion that RRA is not participative, and if it's not PRA then it's exploitative), the need still often exists for 'outsiders' to gain a clearer understanding of local conditions and processes without elevating expectations that follow-on activities will be immediately forthcoming.

2 Originally a navigational expression, triangulation refers to the process of using multiple measurements (i.e. at least three) to determine one's position. The skill of the navigator, or in this case the researcher, in constructing the boundary lines determines the resulting degree of accuracy in locating oneself within the 'triangle.' In social science research, triangulations can be constructed with regard to several dimensions of data (Denzin, 1978). In this study, the principle of triangulation was applied to both different data sources and the methods of data collection used with respect to each source.

3 A large share of this material was drawn from the *RRA Notes* (now *PLA Notes*) and additional manuals published by the International Institute for Environment and Development (IIED), in London (e.g. Gueye and Schoonmaker-Freudenberger, 1991;

McCracken *et al.*, 1988; *RRA Notes* Volumes 13–15), as well as other sources (e.g. *Agricultural Administration* Vol.8(6); Barker, 1979; *Forest, Trees and People Newsletter*, Nos. 13 and 15/16; *IDS Bulletin* Vol. 12(4); Jiggins and de Zeeuw, 1992; Khon Kaen University, 1987; Rhoades, 1982). In recent years a wealth of published material on RRA, PRA and related techniques have emerged (e.g. Kumar, 1993; Pretty *et al.*, 1995; Slocum *et al.*, 1995; World Bank, 1996; cf. Chambers, 1994a, b, c; amongst others). Extensive collections of both published and 'grey' RRA/PRA material are located at the IDS and IIED documentation centres.

4 These included Peace Corps volunteers and field agents of the Co-operative League of the USA (CLUSA), who assist the formally structured Associations Villageoises and other village groups in initiating local economic activities as part of the USAID-funded DHV project (see Bingen *et al.*, 1994).

5 In all but this last incident, the relationship with my driver, who accompanied me throughout the research period, proved invaluable. After familiarizing himself with the research questions during the first several interviews, and with the knowledge that the translated answer should be that of the group responding and not that of the translator, this individual intervened in a number of instances to help, either with the proper translation of the questions, or to make sure that the translated responses were accurate. In addition, he provided a wealth of information on local customs, and shared freely his knowledge of wild fruits and food plants, which we sampled during our long periods of travel.

6 In fact, the very existence of a GV was contested in the majority of villages visited, with the OHVN claiming that such groups existed and the local residents claiming that they did not, or that they had not seen an extension agent in over a year and were uncertain as to their status.

7 A return visit to each of the study sites had been planned in order to report back to the village groups on the research findings, and to allow these groups an opportunity to respond critically to the general conclusion. However, an unfortunate combination of events prevented this phase of the research from being carried out.

8 Mary Jo Arnoldi of the Smithsonian Institute; Juliana Short of the University of Indiana at Bloomington.

Appendix B

1 This version of the *M'Pesoba* system called for crops to be planted in parallel bands laying perpendicular to the *harmattan* winds (Min. de la Production, n.d.), similar to the *couloir* system of agriculture, promoted during the colonial period by the Belgians in the Democratic Republic of the Congo (formerly Zaire) (Ruddle and Manshard, 1981; cf. de Schlippe, 1956).

2 The correspondence of ODRs to the development zones was only approximate, as the ODRs had their own boundaries. In some cases they covered parts of more than one zone, or overlapped with other ODRs within a single zone (e.g. Bingen, 1985).

3 As one of its first administrative acts, the first national government abolished the old colonial *cantons*, which had effectively preserved the traditional political structure of the Malinké *Kefu* (Leynaud and Cisse, 1978), replacing them with a new administrative structure comprised of *regions*, *cerles*, and *arrondissements*. This action reflected the Marxist ideology of the first regime, which called for the destruction of the traditional peasant power base in order to build a new 'collective spirit' at the village level (see Bingen, 1985:16).

Appendix C

1 Originally, the project called for the establishment of AT training centres as a way of promoting AT. These were later abandoned and attention shifted to the creation of demonstration farms as a way to spread improved AT techniques. Both of these

measures are highly reminiscent of an early attempt by the first National Development Plan to use these same approaches to justify lowered investments in the creation of a national extension service (see Jones, 1976). There were no funds in the OHV project specifically allocated for improving extension.

Appendix D

1 These French institutions were part of the Groupement d'Etudes et de Recherches pour le Développement de l'Agronomie Tropicale (GERDAT), and included the Institut Français de Recherches Fruitières Outre-Mer (IFAC), Institut de Recherches Agronomiques Tropicales et des Cultures Vivières (IRAT), Institut de Recherches du Coton et des Textiles Exotiques (IRCT), and the Institut de Recherches pour les Huiles et Oléagineux (IRHO).
2 Although ICRISAT had been present in Mali since 1976, it did not begin its breeding and agronomy programmes in earnest until the 1979 cropping season (ICRISAT, 1979).
3 This trend was exacerbated in the mid-1980s by the adoption of structural adjustment policies, which virtually eliminated governmental support for national agricultural research. By the end of the 1980s, 25 donor agencies were supporting over 50 separate projects, with the IER directly managing less than 5 per cent of the financial resources supporting agricultural research.

References

N.B. Government documents are listed separately following this main set of references.

Adams, A. 1993. 'Food Insecurity in Mali: Exploring the role of the moral economy'. *IDS Bulletin* Vol.24(4):41–51.

Adesina, A.A. 1992. 'Village-Level Studies and Sorghum Technology Development in West Africa: Case study from Mali'. *In* Moock, J. and R. Rhoades (eds) *Diversity, Farmer Knowledge, and Sustainability*. Ithaca: Cornell University Press.

Agriculture and Human Values. 1989. 'Building on Local Agricultural Knowledge'. *Agriculture and Human Values* Vol.VI(3).

—— 1991. 'Indigenous Agricultural Knowledge Systems and Development'. *Agriculture and Human Values* Vol.VIII(1/2).

Ahn, P.M. 1969. *West African Agriculture. Volume 1: West African Soils* (Third Edition). London: Oxford University Press.

Akasaka, M. 1973. 'Ouelésségougou, A Village in Southern Mali, Its Characteristics as Market Settlement'. *Kyoto University African Studies* Vol.VIII:118–49.

Alders, C., B. Haverkort and L. van Veldhuizen (eds). 1993. *Linking with Farmers: Networking for low-external-input and sustainable agriculture*. London: Intermediate Technology Publications.

Allan, W. 1965. *The African Husbandman*. London: Oliver & Boyd.

Altieri, M.A. 1987. *Agroecology: The scientific basis of alternative agriculture*. Boulder: Westview Press.

—— 1990. 'Why Study Traditional Agriculture?' *In* Carroll, C.R., J.H. Vandermeer and P. Rosset (eds) *Agroecology*. New York: McGraw-Hill.

Altieri, M.A. and S.B. Hecht (eds). 1990. *Agroecology and Small Farm Development*. Boca Raton: CRC Press.

Anon. 1992. 'Regenerating West African Soils'. The Soil Management Collaborative Research Support Program. *Newswire* (August) 1992:1.

—— 1993. 'Pest and Pesticide Management Practices and Policies in Africa: Opportunities for success in integrated pest management, Mali (Draft)'. Systems Approach to Regional Income and Sustainable Resource Assistance (SARSA). Washington, DC: USAID.

Agrawal, A. 1995. 'Dismantling the Divide Between Indigenous and Scientific Knowledge'. *Development and Change* Vol.26(3):413–39.

Arce, A. and Long, N. 1992. 'The Dynamics of Knowledge: Interfaces between bureaucrats and peasants'. *In* Long, N. and A. Long (eds) *Battlefields of Knowledge: The interlocking of theory and practice in social research and development*. London: Routledge.

Argyris, C., R. Putman and D.M. Smith. 1985. *Action Science*. San Francisco: Jossey-Bass.

Arnaiz, M.E.O. 1995. 'Farmers' Organisations in the Technology Change Process: An annotated bibliography'. Overseas Development Institute, *Agricultural Administration (Research and Extension) Network Paper 53*. London: ODI.

Ashby, J.A. and L. Sperling. 1995. 'Institutionalizing Participatory, Client-Driven Research and Technology Development in Agriculture'. *Development and Change* Vol.26(4): 753–70.

Atampugre, N. 1993. *Behind the Lines of Stone: The social impact of a soil and water conservation project in the Sahel*. Oxford: Oxfam.

Aubert, G. and B. Newsky. 1949. 'Note on the Vernacular Names of the Soils of the Sudan and Senegal'. *In Proceedings of the First Commonwealth Conference on Tropical and sub-Tropical Soils*, Commonwealth Bureau of Soil Science (1948). Technical Communication No.46. Harpenden: Commonwealth Bureau of Soil Science.

AZANIA. 1989. 'Special Volume on the History of African Agricultural Technology and Field Systems'. *Azania Journal of the British Institute in Eastern Africa* Vol.XXIV.

Bâ, A.H. 1972. *Aspects de la Civilisation Africane*. Paris: Présence African.

Bailleul, P.C. 1981. *Petit Dictionnaire Bambara-Français Français-Bambara*. England: Avebury Publishing Co.

Bandura, A. 1977. *Social Learning Theory*. Englewood Cliffs: Prentice-Hall Inc.

Barker, D. 1979. 'Appropriate Methodology: An example using a traditional African board game to measure farmers' attitudes and environmental images'. *IDS Bulletin* Vol.10(2):37–40.

Barnett, H.G. 1953. *Innovation: The basis of cultural change*. New York: McGraw-Hill Book Co.

Batterbury, S. 1993. 'Roles For Farmers' Knowledge in Africa'. Overseas Development Institute, *Agricultural Administration (Research and Extension) Network Paper 42b*. London: ODI.

—— 1996. 'Planners or Performers? Reflections on Indigenous Dryland Farming in Northern Burkina Faso'. *Agriculture and Human Values* Vol.13(3):12–22.

Batie, S.S. 1983. 'Soil Conservation Policy for the Future'. *In The Farm and Food System in Transition: Emerging policy issues*. Co-operative Extension Service. East Lansing: Michigan State University.

Bebbington, A.J. 1994a. 'Theory and Relevance in Indigenous Agriculture: Knowledge, agency and organization'. *In* Booth, D. (ed.) *Rethinking Social Development; Theory, research, and practice*. Harlow: Longman Scientific and Technical.

—— 1994b. 'Composing Rural Livelihoods: From farming systems to food systems'. *In* Scoones, I. and J. Thompson (eds) *Beyond Farmer First: Rural people's knowledge, agricultural research and extension practice*. London: Intermediate Technology Publications.

Bebbington, A.J., D. Merrill-Sands and J. Farrington. 1994. 'Farmer and Community Organisations in Agricultural Research and Extension: Functions, impacts and questions'. Overseas Development Institute, *Agricultural Administration (Research and Extension) Network Paper 47*. London: ODI.

BECIS (Bureau d'Études de Conseils et d'Interventions au Sahel). 1991. '*Project de Développement de l'Operation Haute Vallée. Rapport Final. Octobre*, 1991'. Bamako: BECIS.

Becker, L.C. 1990. 'The Collapse of the Family Farm in West Africa? Evidence from Mali'. *The Geographical Journal* Vol.156(3):313–22.

Beebe, J. 1994. 'The Concept of the Average Farmer and Putting the Farmer First: The implications of variability for a farming systems approach to research and extension'. *Journal of Farming Systems Research-Extension* Vol.4(3):1–16.

Bell, M. 1979. 'The Exploitation of Indigenous Knowledge, or the Indigenous Exploitation of Knowledge: Whose use of what for what?'. *IDS Bulletin* Vol.10(2):44–50.

Belloncle, G. 1979. *Jeunes Ruraux du Sahel: Une expérience de formation de jeune alphabétisés au Mali*. Paris: Éditions l'Harmattan.

—— 1989. 'Proposals for a New Approach to Extension Services in Africa'. *In* Roberts, N. (ed.) *Agricultural Extension in Africa*. A World Bank Symposium. Washington, DC: The World Bank.

Bennett, J.W. 1976. 'Anticipation, Adaptation, and the Concept of Culture in Anthropology'. *Science* Vol.192(4242):847–53.

Benor, D., J.Q. Harrison and M. Baxter. 1984. *Agricultural Extension: The training and visit system*. Washington, DC: The World Bank.

Bentley, J.W. 1989. 'What Farmers Don't Know Can't Help Them: The strengths and weaknesses of indigenous knowledge in Honduras'. *Agriculture and Human Values* Vol.VI(3):25–31.

—— 1992. 'Alternatives to Pesticides in Central America: Applied studies of local knowledge'. *Culture and Agriculture* Fall 1992 Number 44:10–13.

—— 1994. 'Facts, Fantasies, and Failures of Farmer Participatory Research'. *Agriculture and Human Values* Vol.11(2/3):140–50.

Berger, P.L. and T. Luckmann. 1966. *The Social Construction of Reality: A treatise in the sociology of knowledge*. New York: Anchor Books, Doubleday.

Bergeret, A. with J.C. Ribot. 1990. *L'Arbre Nourricier en Pays Sahélien*. Paris: Editions de la Maison des Sciences de l'Homme.

Bernsten, R.H., A. Rochim, H. Malian and I. Basa. 1980. 'A Methodology for Constructing an Agro-Economic Profile of Cropping Systems Sites'. Manual prepared for the CRIA Cropping Systems Economics Training Program, June 2–14, 1980. Mimeo.

Berry, J.W. and S.H. Irvine. 1986. '*Bricolage*: Savages do it daily'. *In* Sternberg, R.J. and R.K. Wagner (eds) *Practical Intelligence: Nature and origins of competence in the everyday world.* Cambridge: Cambridge University Press.

Berry, J.W., S.H. Irvine and E.B. Hunt (eds). 1988. *Indigenous Cognition: Functioning in cultural context.* NATO ASI Series D: Behavior and Social Sciences No. 41. Dordrecht: Martinus Nijhoff Publishers.

Berry, S.S. 1984. 'The Food Crisis and Agrarian Change in Africa: A review essay'. *African Studies Review* Vol.27(2):59–112.

—— 1989. 'Social Institutions and Access to Resources'. *Africa* 59(1).

Biggs, S.D. 1989a. 'A Multiple Sources of Innovation Model of Agricultural Research and Technology Promotion'. Overseas Development Institute, *Agricultural Administration (Research and Extension) Network Paper No.6*. London: ODI.

—— 1989b. 'Resource-Poor Farmer Participation in Research: A synthesis of experiences from nine national agricultural research systems'. *OFCOR Comparative Study Paper No.3*. The Hague: ISNAR.

Biggs, S.D. and E.J. Clay. 1981. 'Sources of Innovation in Agricultural Technology'. *World Development* Vol.9(4):321–36.

Bijker, W.E., T.P. Hughes and T.J. Pinch (eds). 1987. *The Social Construction of Technological Systems: New directions in the sociology and history of technology*. Cambridge (USA): MIT Press.

Bille, J.C. 1977. *Etude de la Production Primaire Nette d'un Ecosystème Sahélien*. Paris: Travaux et Documents de l'ORSTOM.

Bingen, R.J. 1985. *Food Production and Rural Development in the Sahel: Lessons from Mali's Operation Riz-Segou*. Westview Special Studies in Social, Political, and Economic Development. Boulder: Westview Press.

Bingen, R.J. and B.M. Simpson. 1997. 'Technology Transfer, Agricultural Development and Democracy in West Africa: Challenges and Opportunities in Mali'. *In* Lee, Y.S. (ed.) *Technology Transfer and Public Policy*. Westport: Quorum Books.

Bingen, J., A. Berthé and B. Simpson. 1993. 'Analysis of Service Delivery Systems to Farmers and Village Associations in the Zone of the *Office de la Haute Vallée du Niger*'. Washington DC: Development Alternative Inc.

—— —— 1994. 'Analysis of Service Delivery Systems to Farmers and Village Associations in the Zone of the *Office de la Haute Vallée du Niger*'. Department of Resource Development Occasional Report. East Lansing: Michigan State University.

Bishop, J. and J. Allen 1989. 'The On-Site Costs of Soil Erosion in Mali'. *Policy Planning and Research Staff Environmental Working Paper No.21*. Environment Department. Washington, DC: World Bank.

Blakie, P., K. Brown, M. Stocking, L. Tang, P. Sillitoe and P. Dixon. 1996. 'Understanding Local Knowledge and the Dynamics of Technical Change in Developing Countries'. Paper prepared for the ODA Natural Resources Systems Programme, Socio-Economic Methodologies Workshop, 29–30 April. London: Overseas Development Institute.

Boffa, J.M., G. Yaméogo, P. Nikiéma and J.B. Taonda. 1996. 'What Future the Shea Tree?'. *Agroforestry Today* Vol.8(4):5–9.

Boughton, D. and B. Henri de Frahan. 1992. 'Agricultural Research Impact Assessment: The case of maize technology adoption in Southern Mali'. Bamako: Institut d'Economie Rurale.

Box, L. 1988. 'Experimenting Cultivators: A method for adaptive agricultural research'. *Sociologia Ruralis* Vol.28(1):62–75.

Bradley, P.N. 1977. 'Vegetation and Environmental Change in the West African Sahel'. *In* O'Keefe, P. and B. Wisner (eds) *Landuse and Development. African Environment Special*

Report 5. London: International African Institute with Environment Training Programme UNEP/IDEP/SIDA.

——1983. 'Peasants, Soils and Classification: An investigation into a vernacular soil typology from the Guidimaka of Mauritania'. *Department of Geography Research Series No.14*. Newcastle Upon Tyne: University of Newcastle Upon Tyne.

Brokensha, D., D.M. Warren and O. Werner (eds). 1980. *Indigenous Knowledge Systems and Development*. Washington, DC: University Press of America.

Brokensha, D. and B.W. Riley. 1986. 'Changes in Uses of Plants in Mbeere, Kenya'. *Journal of Arid Environment* Vol.11(1):75–80.

Brouwer, J., L.K. Fussell and L. Herrmann. 1993. 'Soil and Crop Growth Micro-variability in the West African Semi-Arid Tropics: A possible risk-reducing factor for subsistence farmers'. *Agriculture, Ecosystems and Environments* Vol.45:229–38.

Buckles, D. and H. Perales. 1997. 'Farmer-Based Experimentation with Velvetbean: Innovation within tradition'. Paper presented at the *International Conference on Creativity and Innovation at the Grassroots for Sustainable Natural Resource Management*, Indian Institute of Management/Ahmedabad, India, 11–14 January.

Campbell, C.A. 1996. 'Land Literacy in Australia: Landcare and other new approaches to inquiry and learning for sustainability'. *In* Budelman, A. (ed.) *Agricultural R&D at the Crossroads: Merging systems research and social actor approaches*. Amsterdam: Royal Tropical Institute (KIT).

Campbell, D.T. 1965. 'Variation and Selective Retention in Socio-Cultural Evolution'. *In* Barringer, H.R, G.I Blanksten and R.W. Mack (eds) *Social Change in Developing Areas: A reinterpretation of evolutionary theory*. Cambridge Mass.: Schenkman Publishing Co.

Campilan, D.M. and G. Prain. 1997. 'Exploring Local Knowledge Systems in Rootcrop Agriculture: A step towards user-sensitive research and development'. Paper presented at the *International Conference on Creativity and Innovation at the Grassroots for Sustainable Natural Resource Management*, Indian Institute of Management/Ahmedabad, India, 11–14 January.

Carroll, C.R., J.H. Vandermeer and P.M. Rossett (eds). 1990. *Agroecology*. Biological Resource Management Series. New York: McGraw-Hill.

Carsky, R.J., L. Singh and R. Ndikawa. 1994. 'Suppression of *Striga Hermonthica* on Sorghum Using Cowpea Intercrop'. *Experimental Agriculture* Vol.30:349–58.

Carter, S.E. and H.K. Muwira. 1995. 'Spatial Variability in Soil Fertility Management and Crop Response in Mutoko Communal Area, Zimbabwe'. *Ambio* Vol.24(2):77–84.

Chambers, R. 1983. *Rural Development: Putting the last first*. Harlow: Longman Scientific and Technical.

——1990. 'Microenvironments Unobserved'. *Gatekeeper Series* 22. London: International Institute for Environment and Development.

——1994a. 'The Origins and Practice of Participatory Rural Appraisal'. *World Development* Vol.22(7):953–69.

——1994b. 'Participatory Rural Appraisal (PRA): Analysis of experience'. *World Development* Vol.22(9):1253–68.

——1994c. 'Participatory Rural Appraisal (PRA): Challenges, potentials and paradigm'. *World Development* Vol.22(10):1437–54.

Chambers, R. and B.P. Ghildyal. 1985. 'Agricultural Research for Resource-Poor Farmers: The farmer-first-and-last model'. *Agricultural Administration* Vol.20:1–30.

Chambers, R., A. Pacey and L.A. Thrupp (eds). 1989. *Farmer First: Farmer innovation and agricultural research*. London: Intermediate Technology Publications.

Charreau, C. 1974. 'Soils of Tropical Dry and Dry-Wet Climatic Areas of West Africa and Their Use and Management'. Department of Agronomy, Mimeo 74–26. Ithaca: Cornell University.

Charreau, C. and P. Vidal. 1965. 'Influence de l'Acacia albida Del. sur le Sol, Nutrition Minerale et Rendements des Mils Pennisetum au Senegal'. *l'Agronomie Tropicale* Vol.67:600–625.

Chenevix-Trench, P., M. Tessougué and P. Woodhouse. 1997. 'Land, Water and Local Governance in Mali: Rice production and resource use in the Sourou Valley, Bankass

Cercle'. *Rural Resources Rural Livelihoods Working Paper No. 6*. Institute for Development Policy and Management: University of Manchester.

Chevalier, A. 1900. 'Les Zones et ces Provinces Botanique de C.A.O.F.C.R.' *Acad. Sci.* (Paris) 130:1205–8. *In* Lawson, G.W. (ed.) *Plant Ecology in West Africa: Systems and processes* 1986. Chichester: John Wiley and Sons.

Childe, V.G. 1956. *Society and Knowledge*. London: George Allen and Unwin Ltd.

CGIAR (Consultative Group on International Agricultural Research). 1995. *Annual Report 1994–95*. Washington, DC: CGIAR.

CILSS (Comité Inter-états pour la Lutte contre la Sécheresse au Sahel)/Club du Sahel and Ministry of Rural Development. 1978. 'Integrated Rural Development Projects and Improvement of Agricultural Production Systems'. Bamako Seminar Synthesis, February 20 – March 1, 1978. Bamako: Club du Sahel and Ministry of Rural Development.

CIMMYT (Centro Internacional de Mejoramiento de Maiz y Trigo). 1980. *Planning Technologies Appropriate to Farmers: Concepts and procedures*. London: CIMMYT.

CIRAD (Centre de Coopération Internationale en Recherche Agronomique pour le Développement). 1980. *Memento de l'Agronome*. Paris: Ministère de la Cooperation.

Cleeremans, A. 1993. *Mechanisms of Implicit Learning: Connectionist models of sequence processing*. Cambridge Mass.: Bradford Books, MIT Press.

Cochemé, J. and P. Franquin. 1967. 'An Agroclimatological Survey of a Semi-arid Area in Africa South of the Sahara'. The World Meteorological Organization, Interagency Project on Agroclimatology. *Technical Note No.86*. Geneva: FAO/UNESCO/WMO.

Conable, B.B. 1991. 'Africa's Development and Destiny'. Address to the 27th Session of the Organization of African Unity (OAU), Abuja, Nigeria, June 4, 1991.

Coughenour, C.M. 1984. 'Social Ecology and Agriculture'. *Rural Sociology* Vol.49(1): 1–22.

Coughenour, C.M. and S.M. Nazhat. 1985. 'Recent Changes in Villages and Rainfall Agriculture in Northern Central Kordofan: Communication process and constraints'. *Report No.4 INTSORMIL CRSP*. Dept. of Sociology. Lexington: University of Kentucky.

Coulibaly, B. 1988. 'Tests en Milieu Paysan: Expériences, Observations et Impressions des Paysans. (rappot d'une enquête d'évaluation, 'feed-back' *des paysans)* des zones Nord et Sud de l'OHV Campagne 87/88'. Bamako: DRSPR (Volet OHV).

Coulibaly, O.N. 1987. 'Factors Affecting Adoption of Agricultural Technologies by Small Farmers in sub-Saharan Africa: The case of new varieties of Cowpeas around the agricultural research station of Cinzana, Mali'. Masters Thesis. Department of Agricultural Economics. East Lansing: Michigan State University.

Creevey, L.E. (ed.). 1986. *Women Farmers in Africa: Rural development in Mali and the Sahel*. Syracuse: Syracuse University Press.

Cross, N. and R. Barker (eds). 1991. *At the Desert's Edge: Oral Histories from the Sahel*. London: Panos Publication Ltd.

CTA (Technical Centre for Agricultural and Rural Cooperation). 1996. 'Review and Assessment of Rural Development Programmes in Africa: Proceedings of an international Workshop 17–21 January 1994, Arusha, Tanzania'. Wageningen: CTA.

Dabin, B. 1951. '*Contribution a l'Étude des Sols du Delta Central Nigérien*'. *l'Agronomie Tropicale* Vol.VI (No.11–12):606–37.

Davies, S. 1996. *Adaptable Livelihoods: Coping with food insecurity in the Malian Sahel*. London: Macmillian Press.

Davy, E.G., F. Mattei and S.I. Solomon. 1976. 'An Evaluation of Climate and Water Resource for Development of Agriculture in the Sudano-Sahelian zone of West Africa'. World Meteorological Organization, *Special Report No.9*. Geneva: UNEP/WMO.

Day, J.C. 1989. 'Soil and Water Management in West Africa: An economic analysis'. Resources and Technology Division, Economic Research Service, United States Department of Agriculture. *Staff Report No. AGES 89–35*. Washington DC: USDA.

de Boef, W., K. Amanor, K. Wellard with A. Bebbington (eds). 1993. *Cultivating Knowledge: Genetic diversity, farmer experimentation and crop research*. London: Intermediate Technology Publications.

de Schlippe, R. 1956. *Shifting Cultivation in Africa: The Zande system of Agriculture.* London: Routledge and Kegan Paul.
Defoer, T. S. Kanté, T. Hilhorst and H. de Groote. 1996. 'Towards More Sustainable Soil Fertility Management'. Overseas Development Institute, Agriculture Research and Extension Network *Paper No.63.* London: ODI.
Denzin, N.K. 1978. *The Research Act: A theoretical introduction to sociological methods* (2nd ed.). New York: McGraw-Hill.
Diallo, A. 1990. 'Participation Paysanne et Dévelopment Rural: Le cas de l'Opération Haute Vallée (OHV) du Niger au Mali'. Thèse doctorat. Department des Sciences Sociales Appliquées aux Developpment. Laboratoire d'Anthropologie et de Sociologie (LAST). Tours: Universite Francois-Rabelais de Tours.
Diallo, D. 1991. 'Etude des Classifications Paysannes des Sols au Mali: Cas du Djitoumou'. *Atelier sur les Approches et Priorités de Recherches en Appiu a la Lute Communicutaire Contre la Desertification, 9–11 Septembre,* 1991. Dakar: Centre de Recherche pour le Développment International, Bureau Regional pour l'Afrique Centrale et Occidentale.
Diallo, D. and D. Keita. 1992. 'Projet de Recherche sur les Classifications Paysannes des sols au Mali'. Sythèse des Résultats Obtenus dans la Zone Agroécologique du Djitoumou (HBN2). Présentée au Comité Technique Régional de la Recherche Agronomique, Bamako 24–28 Mars, 1992. Bamako: Laboratoire d'Agropédologie, et Institut Polytechnique Rural de Katibougou.
Diarra, N. 1975. 'Le Jardinage Urban et Suburbain au Mali: Le sac de Bamako'. *Journal D'Agriculture Tropicale et de Botanique Appliquée* XXII(10–12):359–64.
Diouf, J. 1990. 'The Challenge of Agricultural Development in Africa'. *CTA Annual Report 1990.* Wageningen: Technical Centre for Agricultural and Rural Cooperation (CTA).
Diop, M. 1971. *Histoire des Classes Sociales dans l'Agrique de l'Ouest: 1. Le Mali. Testes à l'Appui.* Paris: François Maspero.
Dissanayake, W. 1986. 'Understanding the Role of the Environment in Knowledge Generation and Use: A plea for a hermeneutical approach'. *In* Beal, G.M., W. Dissanayake and S. Konoshima (eds) *Knowledge Generation, Exchange, and Utilization.* Boulder: Westview Press.
Djibo, H., C. Coulibaly, P. Marko and J.T. Thomson. 1991. 'Decentralization, Governance and Management of Renewable Natural Resources: Local options in the Republic of Mali. Volume III, October 1991. Final Report'. *Studies on Decentralization in the Sahel. OECD Contract No. 90/52.* Burlington: Associates in Rural Development, Inc.
Doenges, C.E. 1988. 'The Sahel as a Place: The evolution of an image'. *Department of Geography Discussion Paper No.94.* Syracuse: Syracuse University.
Dommen, A.J. 1988. *Innovation in African Agriculture.* Boulder: Westview Press.
——1989. 'A Rationale for African Low-resource Agricultural in Terms of Economic Theory'. *In* Warren, D.M., L. J. Slikkerveer and S.O. Titilola (eds) *Indigenous Knowledge Systems: Implications for agricultural and international development. Studies in Technologies and Social Change No.11.* Ames: Iowa State University.
Drinkwater, M. 1992. 'Visible Actors and Visible Researchers: Critical hermeneutics in an actor-oriented perspective'. *Sociologia Ruralis* 32(4):367–88.
——1994. 'Developing Interaction and Understanding: RRA and farmer research groups in Zambia'. *In* Scoones, I. and J. Thompson (eds) *Beyond Farmer First: Rural people's knowledge, agricultural research and extension practice.* London: Intermediate Technology Publications.
Eicher, C.K. 1989. 'Sustainable Institutions for African Agricultural Development'. *ISNAR Working Paper No. 19.* The Hague: ISNAR.
Eponou, T. 1996. 'Partners in Technology Generation and Transfer: Linkages between research and farmers' organization in three selected African countries'. The Hague: ISNAR.
Fairhead, J. 1993. 'Representing Knowledge: The "new farmer" in research fashions'. *In* Pottier, J. (ed.) *Practising Development: Social science perspectives.* London: Routledge.

Fairhead, J. and M. Leach. 1994. 'Declarations of Difference'. *In* Scoones, I. and J. Thompson (eds) *Beyond Farmer First: Rural people's knowledge, agricultural research and extension practice*. London: Intermediate Technology Publications.

——1996. *Misreading the African Landscape: Society and ecology in a forest-savanna mosaic*. Cambridge: Cambridge University Press.

Falconer, J. 1990. 'The Major Significance of 'Minor' Forest Products'. FAO Community Forestry Note 6. Rome: FAO.

Fals-Borda, O. and M.A. Rahman (eds). 1991. *Action and Knowledge: Breaking the monopoly with participatory action-research*. New York: The Apex Press.

FAO (Food and Agriculture Organization of the United Nations). 1995. 'Understanding Farmers' Communication Networks: An experience in the Philippines'. *Communication for Development Case-Study 14*. Rome: FAO.

Farmer, G. and T.M.L. Wigley. 1985. *Climatic Trends for Tropical Africa: A research report for the Overseas Development Administration*. Climatic Research Unit, School of Environmental Sciences. East Anglia: University of East Anglia.

Fleuret, A. 1986. 'Indigenous Responses to Drought in Sub-Saharan Africa'. *Disasters* Vol.10(3):224–9.

Foltz, J. 1991. 'Mali Livestock Sector II Project, New Livestock Project Design and Livestock Marketing Analysis'. International Agricultural Programs. Madison: University of Wisconsin.

Foster, M. 1972. 'An Introduction to the Theory and Practice of Action Research in Work Organizations'. *Human Relations* Vol.25(6):529–56.

Foster, G.M. 1973. *Traditional Societies and Technological Change*. Second Edition. New York: Harper and Row Publishers.

Francis, C.A. 1986. *Multiple Cropping Systems*. New York: Macmillan Publishing Co.

Francis, C.A. and J.H. Sanders. 1978. 'Economic Analysis of Bean and Maize Systems: Monoculture versus associated cropping'. *Field Crops Research* 1:319–35.

Franke, R.W. and B.H. Chasin. 1980. *Seeds of Famine: Ecological destruction and the development of dilemma in the West African Sahel*. Landmark Studies. Montclair: Allanheld, Osmun and Co. Publishers.

Freire, P. 1970. *Pedagogy of the Oppressed*. New York: Continuum.

Gallais, J. 1967. *Le Delta Intérieur du Niger Vol.1*. Dakar: IFAN.

Gamser, M.S. 1988. 'Innovation, Technical Assistance and Development: The importance of technology users'. *World Development* Vol.16(6):711–21.

Giddens, A. 1979. *Central Problems in Social Theory: Action, structure and contradiction in social analysis*. London: Macmillan Press.

Gladwin, C.H. 1989. 'Indigenous Knowledge Systems, the Cognitive Revolution. and Agricultural Decision Making'. *Agriculture and Human Values* Vol.VI(3):32–41.

Gladwin, H. and M. Murtaugh. 1980. 'The Attentive-Preattentive Distinction in Agricultural Decision Making'. *In* Barlett, P.F. (ed.) *Agricultural Decision Making: Anthropological contributions to rural development*. New York: Academic Press.

Glantz, M. (ed.). 1987. *The Crisis in African Agriculture*. London: Zed Books.

Glaser, B.G. and A.L. Strauss. 1967. *The Discovery of Grounded Theory: Strategies for qualitative research*. Chicago: Aldine.

Gliessman, S. 1990. *Agroecology: Researching the ecological basis for sustainable agriculture*. New York: Springer-Verlag.

Gliessman, S.R., R. Garcia and M.A. Amador. 1981. 'The Ecological Basis for the Application of Traditional Agricultural Technology in the Management of Tropical Agro-Ecosystems'. *Agro-Ecosystems* 7:173–85.

Gnägi, A. 1991. '*Nous Avons Perdu le Respect, Mais la Pitié n'est pas Venu dans Nos Coeurs: Prozesse des sozialen wandels und soziokulturelle heterogenität im arrondissement Ouélessébougou, Mali*'. Berne: Institut für Ethnologie.

——1992. '*Elaboration Participative de Technologies: Développement s'une méthode à partir de l'example de la technologie apicole locale dans l'arrondissement de Ouelessébougou, Mali*'. Berne: Institut d'Ethnologie.

Granovetter, M.S. 1973. 'The Strength of Weak Ties'. *American Journal of Sociology* 78:1360–80.

Grigsby, B. 1989. 'Lending, Borrowing and Women's Social Organizations in Rural Mali'. Department of Forest Resources. University of Idaho. Moscow: Consortium for International Development/Women in Development.

Gritzner, J.A. 1988. *The West African Sahel: Human agency and environmental change. Geography Research Paper No.226*. Chicago: University of Chicago.

Grosz-Ngaté, M. 1986. *Bambara Men and Women and the Reproduction of Social Life in Sana Province, Mali.* Doctoral Dissertation. Department of Anthropology. East Lansing: Michigan State University.

—— 1989. 'Hidden Meanings: Exploration into a Bamanan Construction of Gender'. *Ethnology* Vol.28(3):167–82.

Gubbels, P. 1988. 'Peasant Farmer Agricultural Self-Development'. *ILEIA* 4(3):11–14.

—— 1992. 'Farmer-First Research: Populist pipedream or practical paradigm. Prospects for Indigenous Agricultural Development in West Africa'. Masters Thesis. East Anglia: University of East Anglia.

—— 1997. 'Strengthening Community Capacity for Sustainable Agriculture'. *In* van Veldhuizen, L., A. Waters-Bayer, R. Ramirez, D. Johnson and J. Thompson (eds) *Farmers' Research in Practice: Lessons from the field.* London: Intermediate Technology Publications.

Gueye, B. and K. Schoonmaker Freudenberger. 1991. '*Introduction à la Méthod Accélérée de Recherche Participative* (MARP): *Quelques notes pour appuyer une formation practique*'. London: International Institute for Environment and Development.

Guillet, D., L. Furbee, J. Sandor and R. Benfer. 1995. 'A Methodology for Combining Cognitive and Behavioral Research: The Lari soils project'. *In* Warren, D.M., L.J. Slikkerveer and D. Brokensha (eds) *The Cultural Dimension of Development: Indigenous knowledge systems.* London: Intermediate Technology Publications.

Guinko, S. and L.J. Pasgo. 1992. 'Harvesting and Marketing of Edible Products from Local Woody Species in Zitenga, Burkina Faso'. *Unasylva* 168, Vol.43:16–19.

Gupta, A.K. 1990. 'Inventory of Peasants, Innovations for Sustainable Development: An annotated bibliography'. Centre for Management in Agriculture. Ahmedabad: Indian Institute of Management.

—— 1995. 'Survival Under Stress: Socioeconomic perspectives on farmers' innovations and risk adjustment'. *In* Warren, D.M., L.J. Slikkerveer and D. Brokensha (eds) *The Cultural Dimension of Development: Indigenous knowledge systems.* London: Intermediate Technology Publications.

Gurvitch, G. 1971. *The Social Frameworks of Knowledge.* Oxford: Basil Blackwell.

Habermas, J. 1976. *Communication and the Evolution of Society.* Boston: Beacon Press.

Hammersley, M. and P. Atkinson. 1983. *Ethnography: Principles in practice.* New York: Tavistock Publications.

Haray, F. and R.G. Havelock. 1971. 'Anatomy of a Communication Arc'. *Human Relations* 25(5):413–26.

Harlan, J.R. and J. Pasquereau. 1969. '*Décure* Agriculture in Mali'. *Economic Botany* Vol.23(1):70–74.

Harris, F. 1996. 'Intensification of Agriculture in Semi-Arid Areas: Lessons from Kano Close-Settled Zone Nigeria'. *International Institute for Environment and Development Gatekeeper Series No. 59.* London: IIED.

Harrison, P. 1987. *The Greening of Africa: Breaking through the battle for land and food.* New York: Viking/Penguin, Inc.

Hatch, J. 1976. 'The Corn Farmers of Motupe: A study of traditional farming practices in Northern Coastal Peru'. Monograph No.1. University of Wisconsin-Madison. Madison: Land Tenure Center. *In* Chambers, R. *Rural Development: Putting the last first*, 1983. Harlow: Longman Scientific and Technical.

Havelock, R.G. 1973. *Planning for Innovation Through Dissemination and Utilization of Knowledge.* Center for Research on Utilization of Scientific Knowledge, Institute for Social Research. Ann Arbor: University of Michigan.

Haverkort, B., J. van der Kamp and A. Waters-Bayer (eds). 1991. *Joining Farmers' Experiments*. London: Intermediate Technology Publications.

Haverkort, B., W. Hiemstra, D. Millar and S. Rist (eds). 1996. 'Agri-Culture and Cosmovision'. COMPAS/ETC. Leusden: ETC Netherlands.

Hayward, D. and J. Oguntoyinbo. 1987. *The Climatology of West Africa*. London: Hutchinson.

Heinrich, G. 1993. 'Strengthening Farmer Participation Through Groups: Experiences and lessons from Botswana'. *OFCOR Discussion Paper 3*. The Hague: ISNAR.

Henning, R.K. 1989. 'Programme Spécial Energie Mali: Production d'huile de Pourghere comme carburant'. *Programme Spécial Energie*. Bamako: GTz and Direction National de Hydaulique et de l'Energie.

—— 1992. '*Promotion de la Lutte Contre l'Erosion au Sahel Grace à la Production et à l'Utilisation d'Huile Végétale comme Carburant*'. Eschborn: GTz.

Herrera, A.O. 1981. 'The Generation of Technologies in Rural Areas'. *World Development* Vol.9:21–35.

Hiemstra, W., C. Reijntes and E. van der Werf (eds). 1992. *Let Farmers Judge: Experiences in assessing agricultural innovation*. London: Intermediate Technology Publications.

Hills, T.L. and R.E. Randall (eds). 1968. 'The Ecology of the Forest/Savanna Boundary (Proceedings of the I.G.U Humid Tropics Commission Symposium, Venezuela, 1964)'. McGill University Savanna Research Project, *Savanna Research Series No.13*. Department of Geography. Montreal: McGill University.

Holtzman, J.S. 1986. 'Rapid Reconnaissance Guidelines for Agricultural Marketing and Food Systems Research in Developing Countries'. MSU International Development Papers. *Working Paper No. 30*. Department of Agricultural Economics. East Lansing: Michigan State University.

Holzner, B. and J.H. Marx. 1979. *Knowledge Application: The knowledge system in society*. Boston: Allyn and Bacon, Inc.

Hopkins, B. 1965. *Forest and Savanna: An introduction to tropical plant ecology with special reference to West Africa*. London: Heinemann.

Hoskins, M. 1994. 'Supporting Farmer Extension and Research'. *Forest, Trees and People Newsletter* No.23:4–8.

Hulme, M. 1996. 'Climate Change within the Period of Meteorological Records'. *In* Adams, W.M., A.S. Goudie and A.R. Orme (eds) *The Physical Geography of Africa*. Oxford University Press: Oxford.

Hunt, D. 1991. 'Farming System and Household Economy as Framework for Prioritising and Appraising Technical Research: A critical appraisal of current approaches'. *In* Haswell, M. and D. Hunt (eds) *Rural Households in Emerging Societies*. Oxford: Berg Publication.

IBRD (International Bank for Reconstruction and Development). 1970. 'Economic Development in Mali: Evolution, Problems and Prospects. Volume II Notes on Agriculture, Livestock and Fisheries'. *Report No. AW-14a*. Washington D.C.: IBRD.

ICRISAT (International Crop Research Institute for the Semi-Arid Tropics). 1978. *ICRISAT Annual Report 1978*. Ouagadougou, Republic of Upper Volta, Ministry of Rural Development. In Jaeger, W. *Agricultural Mechanization: The economics of animal draft power in West Africa* (1986). Boulder: Westview Press.

—— 1979. *Report of the 1979 Season: Cooperative Program*. Bamako: ICRISAT.

—— 1985. *ICRISAT Annual Report 1984*. Patancheru: ICRISAT.

—— 1989. *ICRISAT Annual Report 1988*. Patancheru: ICRISAT.

IDS. 1979. 'Rural Development: Whose Knowledge Counts?'. *IDS Bulletin* Vol.10(2).

—— 1981. 'Rapid Rural Appraisal'. *IDS Bulletin* Vol.12(4).

IFAD (International Fund for Agricultural Development). 1993. 'The State of the World Rural Poverty: A profile of Africa'. Rome: IFAD.

ILEIA. 1997. 'Fighting Back with IPM'. *ILEIA Newsletter* Vol.13(4).

Imperato, P.J. and E.M. Imperato. 1982. *Mali: A handbook of Historical Statistics*. Boston: G.K. Hall & Co.

Inglis, J.T. (ed.). 1993. *Traditional Ecological Knowledge: Concepts and cases.* International Program on Traditional Ecological Knowledge and International Development Research Centre. Ottawa: Canadian Museum of Nature.

Innis, D. 1997. *Intercropping and the Scientific Basis of Traditional Agriculture.* London: Intermediate Technology Publications.

IRAT (Institut de Recherches Agronomiques Tropicales et des Cultures Vivieres). 1974. '*Bilan Techniques et Financier des Recherches en Matière d'Agronomie Generale et de Cultures Vivières Conduites par l'IRAT de 1962 a 1974 au Mali: Proposition de Programme Pluriannuel de Recherches*'. Bamako: IRAT.

—— 1975. '*Fiches Techniques*'. Bamako: IRAT.

ISNAR (International Service for National Agricultural Research). 1990. '*Analyse du Système national de Recherche Agronomique du Mali*'. Rapport au Minstère de l'Agriculture, République du Mali. ISNAR R46f. The Hague: ISNAR.

IUCN (International Union for Conservation of Nature). 1989. *The IUCN Sahel Studies: 1989.* Nairobi: IUCN.

Jaeger, W.K. 1986. *Agricultural Mechanization: The economics of animal draft power in West Africa.* Westview Special Studies in Agricultural Science and Policy. Boulder: Westview Press.

Jago, N.D., A.R. Kremer and C. West. 1993. 'Pesticides on Millet in Mali'. Natural Resources Institute. *Bulletin 50.* Chatham Maritime: NRI.

Jiggins, J. and H. de Zeeuw. 1992. 'Participatory Technology Development in Practices: Process and methods'. *In* Reintjes, C., B. Haverkort and A. Waters-Bayer (eds) *Farming For the Future: An introduction to low-external input and sustainable agriculture.* London: Macmillan Press.

Johannes, R.E. (ed.). 1989. *Traditional Ecological Knowledge: A collection of essays.* Gland and Cambridge: IUCN.

Johnson, A.W. 1972. 'Individuality and Experimentation in Traditional Agriculture'. *Human Ecology* Vol.1(2):149–59.

Jones, B. 1938. 'Desiccation and the West African Colonies'. *The Geographical Journal* Vol.XCI(5):401–23.

Jones, W.I. 1976. *Planning and Economic Policy in Mali.* Thèse No.230 Université de Genève, Institut Universitaire de Hautes Etudes Internationales. Washington, DC: Three Continents Press.

Jones, M.J. and A. Wild. 1975. 'Soils of West African Savanna'. *Technical Communication No.55.* Harpenden: Commonwealth Bureau of Soils.

Jones, N. and J.H. Miller. (no date). '*Jatropha curacas*: A Multipurpose species for problematic Sites'. *ASTAG Technical Papers. Land Resources Series No.1.* Asia Technical Department Agricultural Division. Washington, DC: The World Bank.

Josens, G. 1983. 'The Soil Fauna of Tropical Savannas. III The Termites'. *In* Bourlière, F. (ed.) *Ecosystems of the World Vol.14: Tropical Savannas.* Amsterdam: Elsevier Scientific Publishing Co.

Jungerius, P.D. 1985. 'Perceptions and Use of the Physical Environment in Peasant Societies'. The 16th Norma Wilkinson Memorial Lecture. *Reading Geographical Papers No.93.* Reading: University of Reading.

Kagbo, R.B. 1986. *Some Observations on the OHVN On-Farm Recherche and Extension Programs (Background information for preparing the Project's Phase II).* Bamako: OHV.

—— 1987. *Recommendations aux Paysans de l'OHV (en rapport avec les résultats des programmes d'essais en mileau paysan).* Bamako: OHV.

—— 1988. *Bénéfices Attendus de l'Adoption de Certaines Innovations dans la Zone OHV.* Bamako: OHVN.

Kagbo, R.B. and A. Diarra, 1988. *Résultats des Essais Realisés en Zone OHV en* 1987. Bamako: OHVN.

Kater, L.J.M., S. Kante and A. Budelman. 1992. '*Karité (Vitellaria paradoxa)* and *Néré (Parkia biglobosa)* Associated with Crops in South Mali'. *Agroforestry Systems* 18:89–105.

Kates, R.W. 1981. 'Drought Impact in the Sahelian-Sudanic Zone of West Africa: A comparative analysis of 1910–15 and 1968–74'. *Background Paper No.2*. Center for Technology, Environment and Development. Worcester: Clark University.

Katz, C. 1991. 'Sow What You Know: The struggle for social reproduction in rural Sudan'. *Annals of the Association of American Geographers* Vol.81(3):488–515.

Kelly, G.A. 1955. *The Psychology of Personal Constructs. Volume 1. A Theory of Personality*. New York: W.W. Norton & Co., Inc.

Kessler, J.J. 1992. 'The Influence of *Karité (Vitellaria paradoxa)* and *Néré (Parkia biglobosa)* Trees on Sorghum Production in Burkina Faso'. *Agroforestry Systems* 17:97–118.

Khon Kaen University. 1987. *Proceedings of the 1985 International Conference on Rapid Rural Appraisal*. Khon Kaen, Thailand: Rural Systems Research and Farming Systems Research Projects.

King, G.R. 1986. 'Economics of Farming Systems Study, Operation Haute Vallée, Mali'. Washington, DC: Checchi and Company Consulting, Inc.

Kingsbury, D., C. McCorkle and A. Fessenden. 1994. 'Sustainable Cropland Management in Mali: Actions, contributing conditions, practices and change'. Washington, DC: USAID.

Klaij, M.C. and W.B. Hoogmoed. 1989. 'Crop Response to Tillage Practices in a Sahelian Soil'. In *Soil, Crop, and Water Management Systems for Rainfed Agriculture in the Sudano-Sahelian Zone: Proceedings of an International Workshop, 7–11 January, 1987, ICRISAT Center, Niamey, Niger*. Patancheru: ICRISAT.

Knight, C.G. 1974. *Ecology and Change: Rural modernization in an African Community*. New York: Academic Press.

—— 1980. 'Ethnoscience and the Africa Farmer: Rationale and strategy'. *In* Brokensha, D., D.M. Warren and O. Werner (eds) *Indigenous Knowledge Systems and Development*. Washington, DC: University Press of America.

Knorr-Cetina, K.D. 1981. *The Manufacture of Knowledge: An essay on the constructivist and contextual nature of science*. Oxford: Pergamon Press.

Koenig, D. 1986a. 'Social Stratification and Labor Allocation in Peanut Farming in the Rural Malian Household'. *African Studies Review* Vol.(3):107–27.

—— 1986b. 'Research for Rural Development: Experiences of an Anthropologist in Rural Mali'. *In* Horowitz, M.M. and T.M. Painter (eds) 'Anthropology and Rural Development in West Africa'. *IDA Monographs in Development Anthropology*. Boulder: Westview Press.

Kowal, J.M. and A.H. Kassam. 1978. *Agricultural Ecology of Savanna: A study of West Africa*. Oxford: Clarendon Press.

Kremer, A. and K. Sidibé. 1991. 'Mali Millet Pest Project (*Projet Pilot Britannique*) Economics Report 1990'. Natural Resources Institute. Chatham Maritime: NRI.

Krings, T. 1987. 'The Advantages and Risks of Ox-Plowing and Monoculture in the Central and Southern Savannas of the Republic of Mali'. *Applied Geography and Development* Vol.30:46–63.

—— 1991. 'Indigenous Agricultural Development and Strategies for Coping with Famine: The case of *Senfou* (*Pomporo*) in Southern Mali (West Africa)'. *Bayreuther Geowissenschaftliche Arbeiten* Vol.15:69–81.

Kruglanski, A.W. 1989. *Lay Epistemics and Human Knowledge*. New York: Plenum.

Kuhn, T.S. 1970. *The Structure of Scientific Revolutions*. 2nd edn. Chicago: University of Chicago Press.

Kumar, K. 1993. *Rapid Appraisal Methods*. World Bank Regional and Sectoral Studies. Washington, DC: The World Bank.

Lachenmann, G. 1986. 'Rural Development in Mali: Destabilization and social organization'. *Quarterly Journal of International Agriculture* Vol.25(3):217–33.

Lamers, J.P.A., P.R. Feil and A. Buerkert. 1995. 'Spatial Crop Growth Variability in Western Niger: The knowledge of farmers and researchers'. *Indigenous Knowledge and Development Monitor* Vol.3(3):17–19.

Lawler, E.E. III. 1985. 'Challenging Traditional Research Assumptions'. *In* Lawler, E.E., A.M. Mohrman, Jr., S.A. Mohrman, G.E. Ledford, Jr., T.G. Cummings and Associates (eds) *Doing Research That Is Useful for Theory and Practice*. San Francisco: Jossey-Bass Publishers.

Lawson, G.W. (ed.). 1986. *Plant Ecology in West Africa: Systems and processes*. Chichester: John Wiley and Sons.

Lebeau, F. 1986. 'Technology Transfer Study *Opération Haute Vallée* Mali: Second Phase'. Washington, DC: Checci and Company Consulting Inc.

Le Moigne, M. and M. Chavatte. 1972. *Etude de l'Evolution des Facteurs de Production Mise en Place Pendant les Dix Dernières Années et de Leurs Effets*. Bamako: CEEMAT/SEAE

Lewin, K. 1946. 'Action Research and Minority Problems'. *Journal of Social Issues* Vol.2: 34–46.

Lewis, J.V.D. 1979. *Descendants and Crops: Two poles of production in a Malian peasant village*. Doctoral Dissertation. New Haven: Yale University.

—— 1981. 'Domestic Labor Intensity and the Incorporation of Malian Peasant Farmers into Localized Descent Groups'. *American Ethnologist* Vol.8(1):53–73.

Leynaud, E. 1962. *La Modernisation Rurale dans la Haute-Vallée du Niger: Mission Leynaud-Roblot*. 4 Volumes. Paris: BDPA

—— 1966. 'Fraternités d'Âge et Sociétés de Culture dans la Haute-Vallée du Niger'. *Cahiers d'Etudes Africaines* Vol.VI(21):41–68.

Leynaud, E. and Y. Cisse. 1978. *Paysans Malinke du Haute Niger (Tradition et développement rural en Afrique Soudanaise)*. Paris: Editions Imprimeries.

Lionberger, H.F. 1959. 'Community Prestige and the Choice of Sources of Farm Information'. *The Public Opinion Quarterly* Vol.23(1):110–18.

Lionberger, H.F. C.J. Yeh and G.D. Copus. 1975. 'Social Change in Communication Structure: Comparative study of farmers in two communities'. *Rural Sociological Society Monograph Number 3*. Morgantown: West Virginia University.

Loevinsohn, M. and B. Simpson. 1998. 'Practising Evolution: A framework for participatory FSR&E'. Paper prepared for the *15th International Symposium of the Association for Farming Systems Research-Extension*, 29 November – 4 December 1998, Pretoria, South Africa.

Loevinsohn, M.E., J. Mugarura and A. Nkusi. 1994. 'Cooperation and Innovation by Farmer Groups: Scale in the development of Rwandan Valley farming systems'. *Agricultural Systems* 46:141–55.

Long, N. 1984. 'Creating Space for Change: A perspective on the sociology of development'. *Sociologia Ruralis* 24(3/4):168–84.

—— 1989. 'Conclusion: Theoretical reflections on actor, structure and interface'. *In* Long, N. (ed.) *Encounters at the Interface: A perspective on social discontinuities in rural development*. Agricultural University: Wageningen.

Long, N. and J.D. van der Ploeg. 1994. 'Heterogeneity, Actor and Structure: Towards a reconstruction of the concept of structure'. *In* Booth, D. (ed.) *Rethinking Social Development: Theory, research and practice*. Essex: Longman Scientific and Technical.

Long, N. and M. Villareal. 1994. 'The Interweaving of Knowledge and Power in Development Interfaces'. *In* Scoones, I. and J. Thompson (eds) *Beyond Farmer First: Rural people's knowledge, agricultural research and extension practice*. London: Intermediate Technology Publications.

Low, A. 1986. 'On-Farm Research and Household Economics'. *In* Moock, J.L. (ed.) *Understanding Africa's Rural Households and Farming Systems*. Westview's Special Studies Series on Africa. Boulder: Westview Press.

Luery, A. 1989. 'Women's Economic Activities and Credit Opportunities in the Operation *Haute Vallée* (OHV) Zone, Mali'. University of Arizona. Tucson: Consortium for International Development/Women in Development.

Luft, J. 1970. *Group Processes: An introduction to group dynamics*. 2nd edn. Palo Alto: Mayfield Publishing Company.

MacDonald, L H. 1986. *Natural Resources Development in the Sahel: The role of the United Nations System*. Tokyo, Japan: The United Nations University.

Maiga, A., B. Teme, B. Coulibaly, L. Diarra, A. Kergna, K. Tigana and J. Winpenny. 1995. 'Structural Adjustment and Sustainable Development in Mali'. *Working Paper 82*. Overseas Development Institute: London.

Maniates, M. 1993. 'Geography and Environmental Literacy'. *Professional Geographer* Vol.45(3):351–54.

Marshall, C. and G.B. Rossman. 1989. *Designing Qualitative Research*. Newbury Park: Sage Publications.

Mathias-Mundy, E. and C. McCorkle. 1989. 'Ethnoveterinary Medicine: An annotated bibliography'. Center for Indigenous Knowledge for Agriculture and Rural Development. *Bibliographies in Technology and Social Change No.6*. Ames: Iowa State University.

Matlon, P.J. 1977. 'The Size Distribution, Structure and Determinants of Personal Income among Farmers in the North of Nigeria'. Doctoral Dissertation. Ithaca: Cornell University.

——1980. 'Local Varieties, Planting Strategies and Early Season Farming Activities in Two Villages of Central Upper Volta'. *West Africa Economics Program Progress Report 2*. Ouagadougou: ICRISAT.

——1990. 'Improving Productivity in Sorghum and Pearl Millet in Semi-Arid Africa'. *Food Research and Institute Studies* Vol.XXII(1):1–43.

Maxwell, S. 1986. 'Farming Systems Research: Hitting a moving target'. *World Development* Vol.14(1):65–77.

Mazur, R.E. 1984. 'Rural Out-Migration and Labor Allocation in Mali'. *In* Goldscheider, C. (ed.) *Rural Migration in Developing Nations: Comparative studies of Korea, Sri Lanka and Mali*. Boulder: Westview Press.

McArthur, D. 1978. 'Information Sources for Farmers from Outside Their Farms'. *Agricultural Administration* 5:275–80.

McCall, M.K. 1995. 'Indigenous Technical Knowledge in Farming Systems of Eastern Africa: A bibliography'. Center for Indigenous Knowledge for Agriculture and Rural Development. *Bibliographies in Technology and Social Change No.9*. Ames: Iowa State University.

McConnell, W. 1993. 'Local Ecological Knowledge and Environmental Management in The Republic of Mali, West Africa'. Masters Thesis. Programme for International Development and Change. Worcester: Clark University.

McCorkle, C.M. and C. Kamité, 1986. 'Farmers Associations Potential Study Operation *Haute Vallée* Mali'. Washington, DC: Checchi and Company Consulting Inc.

McCorkle, C.M., R.H. Brandstetter and G.D. McClure. 1988. 'A Case Study on Farmers Innovations and Communication in Niger'. Communication for Technology Transfer in Agriculture. Washington, DC: Academy for Educational Development.

McCorkle, C.M., E. Mathias and T.W. Schillhorn van Veen (eds). 1996. *Ethnoveterinary Research and Development*. London: Intermediate Technology Publications.

McCorkle, C.M., M. Coulibaly, S. Coulibaly, D. McHugh, G.W. Selleck and D. Sidibe. 1993. 'USAID/Mali Farming Systems Research and Extension Project: Phase II evaluation report'. Washington, DC: Development Alternatives Inc. and Institute for Development Anthropology.

McCracken, J.A., J.N. Pretty and G.R. Conway. 1988. 'An Introduction to Rapid Rural Appraisal for Agricultural Development'. International Institute for Environment and Development, Sustainable Agriculture Programme. London: IIED

McIntire, J. 1981. 'Rice Production in Mali'. *In* Pearson, S.R., J.D. Stryker and C.P. Humphreys (eds) *Rice in West Africa*. Stanford: Stanford University Press.

McTainsh, G.H. and P.H. Walker. 1982. 'Nature and Distribution of Harmattan Dust'. *Zeitschrift für Geomorphologie* 26(4):417–35.

Megahed, H.T. 1970. 'Socialism and Nation-Building in Africa: The case of Mali (1960–1968)'. *Studies on Developing Countries No.36*. Budapest: Center for Afro-Asian Research of the Hungarian Academy of Sciences.

Meillassoux, C. 1960. 'Essai d'Interprétation du Phenomène Économique dans les sociétés traditionnelles d'Auto-Subsistence'. *Cahiers d'Etudes Africaines* I(4):38–67.

Meyer, P., B. Sacko and A. Dembele. 1993. 'Mali Profil de Pauvreté'. Directional Nationale de la Statistique et de l'Informatique: Bamako. *In* Chenevix-Trench, P., M. Tessougué and P. Woodhouse (1997), 'Land, Water and Local Governance in Mali: Rice production and resource use in the Sourou Valley, Bankass Cercle', *Rural Resources Rural Livelihoods Working Paper No. 6*. Institute for Development Policy and Management: University of Manchester.

Miedema, R. and W. van Vuure. 1977. 'The Morphological, Physical and Chemical Properties of Two Mounds of *Macrotermes bellicosus* (*Smeathman*) Compared with Surrounding Soils in Sierra Leone'. *Journal of Soil Science* Vol.28:112–24.

Millar, D. 1996. 'Footprints in the Mud: Re-constructing the diversities in rural people's learning processes'. Doctoral Dissertation. Department of Sociology. Wageningen: Wageningen Agricultural University.

Mills, B. and E. Gilbert. 1990. 'Agricultural Innovation and Technology Testing by Gambian Farmers: Hope for institutionalizing on-farm research in small-country research systems'. *Journal of Farming Systems Research-Extension* Vol.1(2):47–66.

Milne, B.T. 1991. 'Heterogeneity as a Multiscale Characteristic of Landscapes'. *In* Kolasa, J. and S.T.A. Pickett (eds) *Ecological Heterogeneity*. Ecological Studies 86. Springer-Verlag: New York.

Milne, G. 1935. 'Some Suggested Units of Classification and Mapping, Particularly for East African Soils'. *Soil Research* Vol.4(3):183–98.

Miracle, M.P. 1966. *Maize in Tropical Africa*. Madison: University of Wisconsin Press.

—— 1967. *Agriculture in the Congo Basin: Tradition and change in African rural economies*. Madison: University of Wisconsin Press.

Montagne, P. 1985/86. 'Contribution of Indigenous Silviculture to Forestry Development in Rural Areas: Examples from Niger and Mali'. *Rural Africana* Vol.23/24:61–5.

Moormann, F.R. and B.T. Kang. 1978. 'Microvariability of Soils of the Tropics and Its Agronomic Implication with Special Reference to West Africa'. *In* Stelly, M. (ed.) *Diversity of Soils in the Tropics*. American Society of Agronomy, *Special Publication Number 34*. Madison: ASA.

Moris, J. 1989. 'Indigenous versus Introduced Solutions to Food Stress in Africa'. *In* Sahn, D.E. (ed.) *Seasonal Variability in Third World Agriculture: The consequences for food security*. Baltimore: Johns Hopkins Press.

—— 1991. *Extension Alternatives in Tropical Africa*. London: Overseas Development Institute.

Mortimore, M. 1989. *Adapting to Drought: Farmers, famines and desertification in West Africa*. Cambridge: Cambridge University Press.

Moseley, W. 1993. 'Indigenous Agroecological Knowledge Among the Bamabara of *Djitoumou*, Mali: Foundation for a sustainable community'. Masters Thesis. School of Natural Resources. Ann Arbor: University of Michigan.

Mulla, D.J. 1989. 'Soil Spatial Variability and Methods of Analysis'. In *Soil, Crop, and Water Management Systems for Rainfed Agriculture in the Sudano-Sahelian Zone: Proceedings of an International Workshop, 7–11 Jan. 1987, ICRISAT Sahelian Center, Niamey, Niger*. Patancheru: ICRISAT.

Müller, C. 1855. *Geographici Graece Minores Vol 1*. In Hopkins, B. *Forest and Savanna: An introduction to tropical plant ecology with special reference to West Africa* (1965). London: Heinemann. Paris: Didot.

Mulume, S.M. 1997. 'Farmer Research Brigades in Zaire'. *In* van Veldhuizen, L., A. Waters-Bayer, R. Ramirez, D. Johnson and J. Thompson (eds) *Farmers' Research in Practice: Lessons from the field*. London: Intermediate Technology Publications.

Mundy, P. 1991. 'Bibliography of Indigenous Communication'. Unpublished manuscript.

Mundy, P. and J.L. Compton. 1991. 'Indigenous Communication and Indigenous Knowledge'. *Development Communication Report* No.74(3):1–4.

——1995. 'Indigenous Communication and Indigenous Knowledge'. *In* Warren, D.M., L.J. Slikkerveer and D. Brokensha (eds) *The Cultural Dimension of Development: Indigenous knowledge systems*. London: Intermediate Technology Publications.

Mwanthi, M.A. and V.N. Kimani. 1993. 'Agrochemicals: A potential health hazard among Kenya's small-scale farmers'. *In* Forget, G., T. Goodman and A. de Villiers (eds) *Impact of Pesticide Use on Health in Developing Countries*. Ottawa: International Development Research Centre.

Narby, J. and S. Davis. 1983. 'Resource Development and Indigenous Peoples: A comparative bibliography'. Boston: Anthropology Resource Center.

NOAA (National Oceanic and Atmospheric Administration). 1979. 'Weather-Crop Yield Relationships in Drought Prone Countries of Sub-Saharan Africa: Appendix to the final report'. Atmospheric Science Department. University of Missouri-Columbia. Columbia: NOAA/University of Missouri-Columbia.

NRC (National Research Council). 1984a. *Agroforestry in the West African Sahel*. Washington, DC: National Academy Press.

——1984b. *Environmental Change in the West African Sahel*. National Academy Press: Washington, DC.

——1986. *Proceedings of the Conference on Common Property Resource Management*. Washington, DC: National Academy Press.

——1991. *Microlivestock: Little-known small animals with a promising economic future*. Washington, DC: National Academy Press.

——1992. *Neem: A tree for solving global problems*. Washington, DC: National Academy Press.

——1993. *Vetiver: A thin green line against erosion*. Washington, DC: National Academy Press.

——1996. *Lost Crops of Africa: Vol.1 – Grains*. Washington, DC: National Academy Press.

——(forthcoming). *Lost Crops of Africa: Vol.2 – Cultivated and Wild Fruits*. Washington, DC: National Academy Press.

Nazhat, S.M. and C.M. Coughenour. 1987. 'The Communication of Agricultural Information in Sudanese Villages'. *Report No.5 INTSORMIL CRSP*. Department of Sociology. Lexington: University of Kentucky.

N'Diayé. 1970a. *Les Castes au Mali*. Collection 'Hier'. Bamako: Éditions Populaires.

——1970b. *Groupes Ethniques au Mali*. Collection 'Hier'. Bamako: Éditions Populaires.

Neimeyer, R.A. 1985. *The Development of Personal Construct Psychology*. Lincoln: University of Nebraska Press.

Nelson, J. and C. Hall. 1994. 'Movement of Agricultural Information in Rural Areas: A review of the literature'. Unpublished Report. London: Overseas Development Institute.

Niamir, M. 1990. 'Herders' Decision-Making in Natural Resources Management in Arid and Semi-Arid Africa'. Forests, Trees and People Programme. *Community Forestry Note No.4*. Rome: FAO.

Nicholson, S.E. 1982. 'The Sahel: A climatic perspective. Summary'. *Sahel D(82) 187–Summary*. CILSS/Club Du Sahel and Organization for Economic Cooperation and Development.

Niemeijer, D. 1996. 'The Dynamics of African Agricultural History: Is it time for a new development paradigm?' *Development and Change* Vol.27(1):87–110.

Norem, R.H., R. Yoder and Y. Martin. 1989. 'Indigenous Agricultural Knowledge and Gender Issues in Third World Agricultural Development'. *In* Warren, D.M., L.J. Slikkerveer and S.O. Titilola (eds) *Indigeneous Knowledge Systems: Implications for agricultural and international development*. Center for Indigenous Knowledge for Agriculture and Rural Development. *Studies in Technologies and Social Change No.11*. Ames: Iowa State University.

Norman, D.W. 1980. 'The Farming Systems Approach: Relevancy for the small farmer'. *MSU Rural Development Paper 5*. East Lansing: Michigan State University.

Norman, D.W., D.H. Pryor and C.J. Gibbs. 1979. 'Technical Change and the Small Farmer in Hausaland, Northern Nigeria'. *African Rural Economy Program, Paper No. 21*. Department of Agricultural Economics. East Lansing: Michigan State University.

Norman, D., D. Baker, G. Heinrich and F. Worman. 1988. 'Technology Development and Farmer Groups'. Agricultural Technology Improvement Project. Gabarone: Ministry of Agriculture.

Ntare, B.R. and J.H. Williams. 1992. 'Response of Cowpea Cultivars to Planting Pattern and Date of Sowing in Intercrops with Pearl Millet in Niger'. *Experimental Agriculture* Vol.28(1):41–8.

Nye, P.H. and D.J. Greenland. 1960. 'The Soil Under Shifting Cultivation'. *Technical Communication No.51*. Harpenden: Commonwealth Bureau of Soils.

Nyerere, J. 1968. *Freedom and Socialism*. Dar es Salaam: Oxford University Press.

Ohm, H.W. and J. G. Nagy (eds). 1985. *Appropriate Technologies for Farmers in Semi-Arid West Africa*. West Lafayette: Purdue University.

Okali, C., J. Sumberg and J. Farrington. 1994a. *Farmer Participatory Research: Rhetoric and reality*. London: Intermediate Technology Publications.

Okali, C., J. Sumberg and K.C. Reddy. 1994b. 'Unpacking a Technical Package: Flexible messages for dynamic situations'. *Experimental Agriculture* 30:299–310.

Orr, D.W. 1992. *Ecological Literacy: Education and the transition to a postmodern world*. Albany: State University of New York.

ORSTOM (Office de la Recherche Scientifique et Technique Outre-Mer). 1974. République du Mali. Précipitations Journalières de L'Origine des Stations a 1965. Comité Interafricain d'Etudes Hydrauliques. République Française, Ministère de la Coopération. Office de la Recherche Scientifique et Technique Outre-Mer, Sevice Hydrologique.

Pardey, P., J. Roseboom and N. Beintema. 1997. 'Investments in African Agricultural Research'. *World Development* Vol.25(3):409–23.

Pawluk, R.R., A.J. Sandor and J.A. Tabor. 1992. 'The Role of Indigenous Soil Knowledge in Agricultural Development'. *Journal of Soil and Water Conservation* July–August:298–302.

Peace Corps. 1991. '20 Ans au Service du Peuple Malien'. *Rapport Annuel*. Bamako: Peace Corps.

Pieri, C. 1989. *Fertilité des Terres de Savanes: Bilan de trent ans de recherche et de développement agricoles au sud du Sahara*. Ministère de la Coopération et du Développement et le Centre de Coopération Internationale en Recherche Agronomique pour le Développement (CIRAD-IRAT). Paris: Nain Graphic.

Pingali, P., Y. Bigot and H.P. Binswanger. 1987. *The Mechanization and the Evolution of Farming Systems in sub-Saharan Africa*. Baltimore: Johns Hopkins University Press.

PIRT. 1989a. '*Inventaire et Evaluation des Ressources de la Zone OHV: Volume 1 Rapport Technique*'. Bamako: OHVN.

——. 1989b. '*Inventaire et Evaluation des Ressources de la Zone OHV: Volume 2 Annexes.*' Bamako: OHVN.

Platteau, J.P. 1990. 'The Food Crisis in Africa: A comparative structure analysis'. *In* Drèze, J. and A. Sen (eds) *The Political Economy of Hunger. Vol. 2 Famine Prevention*. Oxford: Clarendon Press.

Polanyi, M. 1966. *The Tacit Dimension*. Garden City: Doubleday and Company.

Pottier, J. 1994. 'Agricultural Discourses: Farmer experimentation and agricultural extension'. *In* Scoones, I. and J. Thompson (eds) *Beyond Farmer First: Rural people's knowledge, agricultural research and extension practice*. London: Intermediate Technology Publications.

Potts, M.J., G.A. Watson, R. Sinung-Basuki and N. Gunadi. 1992. 'Farmer Experimentation as a Basis for Cropping Systems Research: A case study involving true potato seed'. *Experimental Agriculture* Vol.29:19–29.

Powell, J.M. and T.O. Williams. 1993. 'Livestock, Nutrient Cycling and Sustainable Agriculture in the West African Sahel'. *International Institute for Environment and Development Gatekeeper Series No.37*. London: IIED.

Prain, G., S. Fujisaka and D.M. Warren (eds). (forthcoming). *Biological and Cultural Diversity: The role of indigenous agricultural experimentation in development*. London: Intermediate Technology Publications.

Pretty, J.N. and R. Chambers. 1993. 'Towards a Learning Paradigm: New professionalism and institutions for agriculture'. *Discussion Paper 334*. Institute of Development Studies: Sussex.

Pretty, J. N., I. Guijt, J. Thompson and I. Scoones. 1995. *A Trainer's Guide for Participatory Learning and Action*. International Institute for Environment and Development, Participatory Methodology Series. London: IIED

Prudencio, C.Y. 1993. 'Ring Management of Soils and Crops in the West African semi-Arid Tropics: The case of the Mossi farming system in Burkina Faso'. *Agriculture, Ecosystems and Environment* 47:237–64.

Rao, M.R. and R.W. Wiley. 1980. 'Evaluation of Yield Stability in Intercropping Studies on Sorghum/Pigeon Pea'. *Experimental Agriculture* 16:105–116.

Radcliffe, E.B., G. Ouedraogo, S. E. Patten, D.W. Ragsdale and P.P. Strzok. 1995. 'Neem in Niger: A new context for a system of indigenous knowledge'. *In* Warren, D.M., L.J. Slikkerveer and D. Brokensha (eds) *The Cultural Dimension of Development: Indigenous knowledge systems*. London: Intermediate Technology Publications.

Ravnborg, H.M. 1992. 'The CGIAR in Transition: Implications for the poor, sustainability and the national research systems'. Overseas Development Institute, *Agricultural Administration (Research and Extension) Network Paper 31*. London: ODI.

Reardon, T. 1997. 'Using Evidence of Household Income Diversification to Inform Study of the Rural Nonfarm Labor Market in Africa'. *World Development* Vol.25(5):735–47.

Reardon, T., C. Delgado and P. Matlon. 1992. 'Determinants and Effects of Income Diversification Amongst Farm Households in Burkina Faso'. *The Journal of Development Studies* Vol.28(2):264–69.

Reardon, T., A.A. Fall, V. Kelly, C. Delgado, P. Matlon, J. Hopkins and O. Badiane. 1994. 'Is Income Diversification Agriculture-Led in the West African Semi-Arid Tropics? The Nature, causes, effects distribution and production linkage of off-farm activities'. *In* Atsain, Wangwe and Drabek (eds) *Economic Policy Experience in Africa: What have we learned?*. Nairobi: African Economic Research Consortium.

Reij, C. 1988. 'The Present State of Soil and Water Conservation in the Sahel'. *Sahel D(89)329*. OCDE/OECD and CILSS.

—— 1991. 'Indigenous Soil and Water Conservation in Africa'. *International Institute for Environment and Development Gatekeeper Series No.27*. London: IIED.

Reij, C., I. Scoones and C. Toulmin (eds). 1996. *Sustaining the Soil: Indigenous soil and water conservation in Africa*. London: Earthscan.

Rhoades, R.E. 1982. *The Art of the Informal Agricultural Survey*. International Potato Center. Lima: CIP.

Rhoades, R.E. and R. Booth. 1982. 'Farmer-Back-to-Farmer: A model for generating acceptable agricultural technology'. *Agricultural Administration* Vol.11:127–37.

Rhoades, R.E. and A. Bebbington. 1995. 'Farmers Who Experiment: An untapped resource for agricultural research and development'. *In* Warren, D.M., L.J. Slikkerveer and D. Brokensha (eds) *The Cultural Dimension of Development: Indigenous knowledge systems*. London: Intermediate Technology Publications.

Richards, P. 1979. 'Community Environmental Knowledge in African Rural Development'. *IDS Bulletin* Vol.10(2):28–36.

—— 1985. *Indigenous Agricultural Revolution*. London: Unwin Hyman Ltd.

—— 1986. 'Coping with Hunger: Hazard and experimentation in an African rice-farming system'. *The London Research Series in Geography No.11*. London: Allen and Unwin.

—— 1989a. 'Farmers Also Experiment: A neglected intellectual resource in African science'. *Discovery and Innovation* Vol.1(1):19–24.

—— 1989b. 'Agriculture as Performance'. *In* Chambers, R., R. Longhurst and A. Pacey (eds) *Farmer First: Farmer innovation and agricultural research*. London: Intermediate Technology Publications.

——1993. 'Cultivation: Knowledge or performance?'. *In* Hobart, M. (ed.) *An Anthropological Critique of Development: The growth of ignorance*. London: Routledge.

——1994. 'Local Knowledge Formation and Validation: The case of rice production in Central Sierra Leone'. *In* Scoones, I. and J. Thompson (eds) *Beyond Farmer First: Rural people's knowledge, agricultural research and extension practice*. London: Intermediate Technology Publications.

Riches, C.R., L.J. Shaxson, J.W.M. Logan and D.C. Munthali. 1993. 'Insect and Parasitic Weed Problems in Southern Malawi and the Use of Farmer Knowledge in the Design of Control Measures'. Overseas Development Institute, *Agricultural (Research and Extension) Administration Network Paper 42a*. London: ODI.

Rodier, J.A. 1982. 'Evaluation of Annual Runoff in Tropical African Sahel'. *Travaux et Documents de l'ORSTOM No.145*. Paris: ORSTOM.

Rogers, E. 1983. *Diffusion of Innovation* (3rd edn). New York: The Free Press.

Röling, N. 1986. 'Extension and the Development of Human Resources: The other tradition in extension education'. *In* Jones, G.E. (ed.) *Investing in Rural Extension: Strategies and goals*. Elsevier: London.

Romanoff, S. 1990. 'On Reducing the Costs of Promoting Local Farmers' Organisations in Agricultural Development Projects'. Paper presented at the 10th Annual ASFSRE Symposium, October 14–17, 1990. East Lansing: Michigan State University.

RONCO. 1985. 'Evaluation Operation Haute Vallée Mali'. Washington, DC: RONCO Consulting Corp.

Roose, E. 1977. 'Erosion et Ruissellement en Afrique de l'Ouest: Vingt années de mesures en petites parcelles expérimentales'. *Travaux et Documents de l'ORSTOM No.78*. Paris: ORSTOM.

Rosenmayr, L. 1988. 'More Than Wisdom: A field study of the old in an African village'. *Journal of Cross-Cultural Gerontology*. Vol.3(1):21–40.

Rosenthal, T.L. and B.J. Zimmerman. 1978. *Social Learning and Cognition: Advances in theory and research*. New York: Academic Press.

Ruddle, K. and W. Manshard. 1981. *Renewable Natural Resources and the Environment*. Dublin: Tycooly International Publishing Ltd.

Ryan, B. and N.C. Gross. 1943. 'The Diffusion of Hybrid Seed Corn in Two Iowa Communities'. *Rural Sociology* Vol 8(1):15–23.

Sall, A. (n.d.). *L'Organisation du Monde Rural du Mali (1910–1988) 1. Evolution et Perspectives 2. Principaux Textes Organization*. Bamako: Edition-Imprimerie du Mali.

SATEC (Aide Technique pour la Coopération et le Développement SODETEG). 1985. '*Etude des Opérations de Développement Rural (ODR) et des Organismes Similaires. Deuxième phase propositions de redressement. Rapport de Synthèse (edition provisoire)*'. Paris: SATEC.

Scarborough, V. (ed.). 1996. 'Farmer-Led Approaches to Extension: Papers presented at a workshop in the Philippines, July 1995'. Overseas Development Institute, *Agricultural Research and Extension Network Papers 59a, b, c*. London: ODI.

Schilling, T., F. Bidinger, O. Coulibaly, E. Smith and B. Teme. 1989. *Final Evaluation: ICRISAT/Mali Project Phase II (688–0226)*. Bamako: USAID.

Schutz, A. and T. Luckmann. 1973. *The Structure of the Life-World*. Evanston: Northwestern University Press.

Scoones, I. and J. Thompson. 1992. 'Beyond Farmer First Rural People's Knowledge, Agricultural Research and Extension Practice: Towards a theoretical framework'. *Beyond Farmer First Overview Paper No.1*. London: IIED.

——(eds). 1994. *Beyond Farmer First: Rural people's knowledge, agricultural research and extension practice*. London: Intermediate Technology Publications.

Scoones, I., M. Melnyk and J.N. Pretty (eds). 1992. *The Hidden Harvest: Wild foods and agricultural systems. A literature review and annotated bibliography*. London: International Institute for Environment and Development.

Scott-Wendt, J., R.G. Chase and L.R. Hossner. 1988. 'Soil Chemical Variability in Sandy Ustalfs in Semiarid Niger, West Africa'. *Soil Science* Vol.145(6):414–19.

SECID (South-East Consortium for International Development). 1987. 'Farming Systems Research and Extension in the Zone of the Opération Haute Vallée du Niger in Mali: A rapid reconnaissance survey'. Mali Farming Systems Research and Extension Project. Chapel Hill: USAID and Institut d'Economie Rurale.

SEDES (Secretariat d'Etat aux Affaires Etrangeres/Société d'Etudes pour le Développement Economique et Social). 1972. *Les Perspectives de Développement A long Terme: A partir du Développement Rural. Tome II. Mali 1968–1986*. Paris: SEDES.

Selener, D. 1997. *Participatory Action Research and Social Change*. The Cornell Participatory Action Research Network. Ithaca: Cornell University.

Sélingué. 1992. *Seminaire: Analyse des systemes de fourniture de service aux paysans et aux associations villageoises. Rapport de Synthese. Novembre 1992*. Bamako: USAID.

Senge, P. 1990. *The Fifth Discipline: The art and practice of the learning organization*. New York: Doubleday Currency.

Seyler, J.R. 1993. 'A Systems Analysis of the Status and Potential of *Acacia albida* in the Peanut Basin of Senegal'. Senegal Agricultural Research II Project. *AID Contract No. 685–0957–C-8004–00*. Dakar: USAID.

Shaikh, A., E. Arnold, K. Christophersen, R. Hagen, J. Tabor and P. Warshall. 1988a. *Opportunities for Sustained Development: Successful natural resource management in the Sahel, Volume 1/Main Report*. Office of Technical Resources and Sahel Office, Africa Bureau. Washington, DC: USAID.

—— 1988b. *Opportunities for Sustained Development: Successful natural resource management in the Sahel, Volume 2/Case Descriptions*. Office of Technical Resources and Sahel Office, Africa Bureau. Washington, DC: USAID.

Sharland, R.W. 1989. 'Indigenous Knowledge and Technical Change in a Subsistence Society: Lessons from the Moru of Sudan'. Overseas Development Institute, Agricultural Administration Unit. *Agricultural Administration (Research and Extension) Network Paper 9*. London: ODI.

—— 1991. 'Trees in the Garden: Interaction between the wild and agricultural domains in practice among the Moru of the southern Sudan'. *Unasylva* 42(164):55–62.

Sikana, P. 1993. 'How Farmers and Scientists See Soils: Mismatched models'. *ILEIA Newsletter* Vol.9(1):15–16.

—— 1994. 'Alternatives to Current Research and Extension Systems: Village research groups in Zambia'. *In* Scoones, I. and J. Thompson (eds), *Beyond Farmer First: Rural people's knowledge, agricultural research and extension practice*. London: Intermediate Technology Publications.

Silverman, R.A. 1994. 'Elema Boru (Artist Profile)'. In *Ethiopia: Traditions of Creativity*. East Lansing: Michigan State University Museum.

Simpson, B.M. 1992. 'Social Organization and the Traditional Management of Natural Resources in Fragile Environments of sub-Saharan Africa'. All-University Research Initiation Grant. Unpublished manuscript. East Lansing: Michigan State University.

—— 1994. 'The Systematic Integration of Knowledge Systems: An underutilized potential for agricultural development in marginal areas'. In *Proceeding of the 13th International Symposium on Systems-Oriented Research in Agriculture and Rural Development, Montpellier, France, November 1994*.

—— 1995. 'Knowledge, Innovation, and Communication: Contribution of the formal and informal systems to agrarian change in the Office de la Haute Vallée du Niger, Mali'. Doctoral Dissertation. Department of Resource Development. East Lansing: Michigan State University.

—— 1998. 'Investing in People: The support of farmer learning, creativity and local social networks in the Projet Pisciculture Familiale, Zäire'. *European Journal of Agricultural Education and Extension*. 5(2): 99–111.

Sivakumar, M.V.K. 1989. 'Agroclimatic Aspects of Rainfed Agriculture in the Sudano-Sahelian Zone'. In *Soil, Crop, and Water Management Systems for Rainfed Agriculture in the Sudano-Sahelian Zone: Proceeding of an International Workshop, 7–11 Jan. 1987, ICRISAT Sahelian Center, Niamey, Niger*. Patancheru: ICRISAT.

Slaybaugh-Mitchell, T. 1995. 'Indigenous Livestock Production and Animal Husbandry: An annotated bibliography'. Center for Indigenous Knowledge for Agriculture and Rural Development. *Bibliographies in Technology and Social Change No.8*. Ames: Iowa State University.

Slocum, R., L. Wichhart, D. Rocheleau and B. Thomas-Slayter (eds). 1995. *Power, Process and Participation: Tools for change*. London: Intermediate Technology Publications.

Sperling, L. 1994. 'The Research and Farmer Organisation Partnership: Methodological reflections on efficiency and empowerment'. Paper presented at the ODI/ISNAR Practitioners' Workshop, 6–8 June, The Hague, The Netherlands.

Sperling, L. and M.E. Loevinsohn. 1993. 'The Dynamics of Adoption: Distribution and mortality of bean varieties among small farmers in Rwanda'. *Agricultural Systems* 41:441–53.

Sperling, L., M.E. Loevinsohn and B. Ntabomvura. 1993. 'Rethinking the Farmer's Role in Plant Breeding: Local bean experts and on-station selection in Rwanda'. *Experimental Agriculture* 29:509–19.

Sperling, L. and J. A. Ashby (forthcoming). 'Moving Participatory Plant Breeding Forward: The next steps'. *In* Collinson, M. (ed.) *History of Farming Systems Research*. Rome: FAO.

SPORE. 1991. *SPORE* No.32:5.

—— 1993. 'Shade Increases Crop Yields'. *SPORE* No.44:10.

Steedman, C., T.E. Davies, M.O. Johnson, and J.W. Sutter, with E.J. Berg, R.J. Bingen and M.J. Morgan. 1976 'Mali: Agricultural Sector Assessment'. *Final Report*. Center for Research on Economic Development, the University of Michigan. Ann Arbor: CRED.

Steentoft, M. 1988. *Flowering Plants in West Africa*. Cambridge: Cambridge University Press.

Steffen, P. 1992. 'The Roles and Limits of the Grain Market in Assuring Food Security in Northeastern Mali: Implications for public policy'. Doctoral Dissertation. Department of Agricultural Economics. East Lansing: Michigan State University.

Steila, D. and T.E. Pond. 1989. *The Geography of Soils: Formation, distribution, and management*. Savage Maryland: Rowman & Littlefield Publishers.

Sternberg, R.J. and R.K. Wagner. (eds). 1986. *Practical Intelligence: Nature and origins of competence in the everyday world*. Cambridge: Cambridge University Press.

Stewart, I. 1986. 'Response Farming: A scientific approach to ending starvation and alleviating poverty in drought zones of Africa'. *In* Moses, Y.T. (ed.) *Proceedings: African Agricultural Development Conference: Technology, ecology and society*. California State Polytechnic University, Pomona, Kellog West Conference Center, May 28–June 1, 1985. Pomona: California State Polytechnic University.

Stolzenback, A. 1994. 'Learning by Improvisation: Farmers' experimentation in Mali'. *In* Scoones, I. and J. Thompson (eds) *Beyond Farmer First: Rural people's knowledge, agricultural research and extension practice*. London: Intermediate Technology Publications.

—— 1997. 'The Craft of Farming and Experimentation'. *In* van Veldhuizen, L., A. Waters-Bayer, R. Ramirez, D. Johnson and J. Thompson (eds) *Farmers' Research in Practice: Lessons from the field*. London: Intermediate Technology Publications.

Stoop, W.A. 1986. 'Agronomic Management of Cereal/Cowpea Cropping Systems for Major Toposequence Land Types in the West African Savanna'. *Field Crops Research* 14:301–19.

—— 1987a. 'Variations in Soil Properties Along Three Toposequences in Burkina Faso and Implications for the Development of Improved Cropping Systems'. *Agriculture Ecosystems and Environment* 19:241–64.

—— 1987b. 'Adaptation of Sorghum/Maize and Sorghum/Pearl Millet Intercrop Systems to the Toposequence Land Types in the North Sudanian Zone of the West African Savanna'. *Field Crops Research* 16:255–72.

Strauss, A. and J. Corbin. 1990 *Basics of Qualitative Research: Grounded theory procedures and techniques*. Newbury Park: Sage Publications.

——1994. 'Grounded Theory Methodology'. *In* Denzin, N.K. and Y.S. Lincoln (eds) *Handbook of Qualitative Research*. Thousand Oaks: Sage Publications.

Sumberg, J. and C. Okali. 1997. *Farmers' Experiments: Creating local knowledge*. Boulder: Lynne Rienner Publishers.

Sundberg, S. 1989. 'OHV Food Consumption and Expenditure Survey: Preliminary results on incomes sources in the OHV'. Unpublished report.

Sutton, J. 1990. *A Thousand Years of East Africa*. Nairobi: British Institute in Eastern Africa.

Swift, J. 1979. 'Notes on Traditional Knowledge, Modern Knowledge and Rural Development'. *IDS Bulletin* Vol.10(2):41–3.

Swindell, K. 1988. 'Agrarian Change and Peri-Urban Fringes in Tropical Africa'. In Rimmer, D. (ed.) *Rural Transformation in Tropical Africa*. Athens: Ohio University Press.

SWI PR/GA (Systemwide Programme on Participatory Research and Gender Analysis). 1997. 'A Global Programme on Participatory Research and Gender Analysis for Technology Development and Organisational Innovation'. Overseas Development Institute, *Agricultural Research and Extension Network Paper No.72*. London: ODI.

Sy, B.S. and M.Y. Bah. 1989. 'Village Associations and Agricultural Extension in the Republic of Mali'. *In* Roberts, N. (ed.) *Agricultural Extension in Africa*. A World Bank Symposium. Washington, DC: The World Bank.

Tamboura, B.B. 1983. 'A Comparative Economic Analysis of Agricultural Production in Three Regions in Mali, West Africa'. Masters Thesis. Graduate College. University of Vermont.

Tangara, M. 1992. 'Family Social Structure, Farm Operation Characteristics and the Adoption of New Technologies for Sustainable Farming Systems in Mali'. Doctoral Dissertation. Department of Sociology. Ames: Iowa State University.

Teme, B, D. Keita and O. Coulibaly. 1993. *Analyse des Relations Entre la Recherche et le Développement ou la Problematique de Transfert des Technologies en Agriculture. Seminaire sur la Liaison recherche-Développement 15–16 Avril*, 1993. Bamako: DRSPR.

Thapa, B., F.L. Sinclair and D.H. Walker. 1995. 'Incorporation of Indigenous Knowledge and Perspectives in Agroforestry Development. Part 2: Case-study on the impact of explicit representation of farmers' knowledge'. *Agroforestry Systems* 30: 249–61.

Thomas, W.I. 1923. *The Unadjusted Girl*. New York: Harper Torch Books. In Holzner, B. and J.H. Marx (1979) *Knowledge Application: The knowledge system in society*. Boston: Allyn and Bacon, Inc.

Thomasson, G.C. 1981. 'Maximizing Participative Planning: Cultural and psychological aspects of user-centered soil resource inventory data preparation and presentation'. In *Soil Resource Inventories and Development Planning: Selected papers from the proceedings of workshops*, Soil Management Support Services, Soil Conservation Services, US Department of Agriculture. *Technical Monograph No.1*. Washington, DC: USDA.

Thompson, J. 1995. 'Participatory Approaches in Government Bureaucracies: Facilitating the process of institutional change'. *World Development* Vol.23(9):1521–54.

Thomson, J.T. 1987. 'Land and Tree Tenure Issues in Three Francophone Sahelian Countries: Niger, Mali, and Burkina Faso'. *In* Raintree, J.B. (ed.) *Land, Trees and Tenure*. Nairobi and Madison: ICRAF and the Land Tenure Center, pp. 211–16.

Thomson, J.T. and C. Coulibaly. 1995. 'Common Property Forest Management Systems in Mali: Resistance and vitality under pressure'. *Unasylva* 180 Vol.46:16–22.

Thomson, J.T., D.H. Feeny and R.J. Oakerson. 1986. 'Institutional Dynamics: The evolution and dissolution of common property resource management'. In National Research Council *Proceedings of the Conference on Common Property Resource Management*. Washington DC: National Academy Press.

Thurston, H.D. 1996. *Slash/Mulch – Sustainable agriculture in the tropics*. London: Intermediate Technology Publications.

Timberlake, L. 1991. *Africa in Crisis: The causes, the cures of environmental bankruptcy*. London: Earthscan Publications.

Tiffen, M., M. Mortimore and F. Gichuki. 1994. *More People, Less Erosion: Environmental recovery in Kenya*. John Wiley and Sons: Chichester.

Toulmin, C. 1986 'Access to Food, Dry Season Strategies and Household Size Amongst the Bambara of Central Mali' IDS Bulletin Vol 17(3):58–66.

Toulmin, C. 1991. 'Staying Together: Household responses to risk and market malfunction in Mali'. *In* Haswell, M. and D. Hunt (eds) *Rural Households in Emerging Societies*. Oxford: Berg Publication.

—— 1992. *Cattle, Women, and Wells: Managing household survival in the Sahel*. Oxford: Clarendon Press.

Towner, G. 1980. *The Architecture of Knowledge*. Washington, DC: University Press of America.

Tripp, R. 1992. 'Expectation and Realities in On-Farm Research'. Paper prepared for the workshop on 'Impacts of On-Farm Research', Harare, Zimbabwe, June 23–26.

UN (United Nations). 1971. 'Integrated Approach to Rural Development in Africa'. Social Development Section of the Economic Commission for Africa. *Social Welfare Services in Africa, No.8*. New York: UN.

USAID (United States Agency for International Development). 1978. 'Opération Haute Vallée (688–0210) Project Paper'. Bamako: USAID.

—— 1982. 'The Opération Haute Vallée Project in Mali is Experiencing Serious Problems'. Regional Inspector General for Audit Abidjan. *Audit Report No. 7–688–82–1*. Bamako: USAID.

—— 1984. 'Farming Systems Research/Extension (688–0232) Project Paper'. Bamako: USAID.

—— 1986. 'Operation Haute Vallée (688–0210) Internal Evaluation, February 1986'. Bamako: USAID.

—— 1988. 'Mali: Development of the Haute Vallée (688–0233) Project Paper'. Bamako: USAID.

—— 1992. 'Strengthening Research Planning and Research on Commodities (SPARC) Project Paper (688–0250)'. Bamako: USAID.

—— 1993. 'Development of *Haute Vallée* Project: Amendment Number One'. Bamako: USAID.

Uphoff, N. 1986. *Local Institutional Development: An analytical sourcebook with cases*. Kumarian Press: West Hartford.

Vanberg, V. 1992. 'Innovation, Cultural Evolution, and Economic Growth'. *In* Witt, V. (ed.) *Explaining Process and Change: Approaches to evolutionary economics*. Ann Arbor: University of Michigan Press.

van den Breemer, J.P.M., C.A. Drijver and L.B. Venema (eds). 1995. *Local Resource Management in Africa*. John Wiley and Sons: Chichester.

van der Ploeg, J.D. 1994. 'Styles of Farming: An introductory note on concepts and methodology'. *In* van der Ploeg, J.D. and A. Long (eds.) *Born From Within: Practice and perspectives of endogenous rural development*. Assen: van Gorcum.

van der Poel, P. and B. Kaya. 1990. '*Synthèse des Activités de Recherche de la DRSPR l'Amenagement Anti-Erosif* (1986–1989)'. Amsterdam: Institut Royal des Tropiques (KIT) and DRSPR (Sikasso).

van der Pol, F. 1992. 'Soil Mining: An unseen contributor to farmer income in southern Mali'. KIT Bulletin 325. Amsterdam: Royal Tropical Institute.

van Shaik, P., M. Gaudreau, J. Lichte, R. Cook and D. Rudisuhle. 1991. 'Evaluation: Farming Systems Research/Extension Project (688–0232)'. Washington, DC: Experience Inc.

van Veldhuizen, L., A. Waters-Bayer, R. Ramirez, D. Johnson and J. Thompson (eds). 1997. *Farmers' Research in Practice: Lessons from the field*. London: Intermediate Technology Publications.

Van Wambeke, A. and R. Dudal. 1978. 'Macrovariability of Soils in the Tropics'. *In* Stelly, M. (ed) *Diversity of Soils in the Tropics*. American Society of Agronomy, *Special Publication Number 34*. Madison: ASA.

Van Westen, A.C.M. and M.C. Klute. 1986. 'From Bamako, with Love: A case study of migrants and their remittances'. *Tijdschrift voor Econ. en Soc. Geografie* Vol.77(1):42–49.

Vandermeer, J. 1989. *The Ecology of Intercropping*. Cambridge: Cambridge University Press.

Vierich, H. 1986. 'Agricultural Production, Social Status, and Intra-Compound Relationships'. *In* Moock, J.L. (ed.) *Understanding Africa's Rural Households and Farming Systems*. Boulder: Westview Press.

Vierich, H.I.D. and W.A. Stoop. 1990. 'Changes in West African Savanna Agriculture in Response to Growing Population and Continuing Low Rainfall'. *Agriculture, Ecosystems and Environment* 31(1990):115–32.

Viguier, P. 1939. *La Riziculture Indigène au Soudan Français*. Paris: Larose.

—— 1952. *'Centre de Colonisation en Terre Sèche de M'Pésoba (Soudan)'*. *L'Agronomie Tropicale* Vol.7(2):180–83.

von Maydell, H.J. 1990. *Trees and Shrubs of the Sahel: Their characteristics and uses*. Deutsche Gesellschaft für Technische Zusammenarbeit (GTZ). Weikersheim: Verlag Josef Magraf.

von Hippel, E.A. 1978. 'Users as Innovators'. *Technology Review* January: 31–9.

Walker, D.H., F.L. Sinclair and R.I. Muetzelfeldt. 1991. 'Formal Representation and Use of Indigenous Ecological Knowledge About Agroforestry: A pilot phase report'. School of Agriculture and Forest Science. Bangor: University of Wales.

Walter, G. 1993. 'Farmers' Use of Validity Cues to Evaluate Reports of Field-Scale Agricultural Research'. *American Journal of Alternative Agriculture* Vol.8(3):107–117.

Warner, K. 1991. 'Shifting Cultivators: Local technical knowledge and natural resource management in the humid tropics'. Forests, Trees and People Programme. *Community Forestry Note No.8*. Rome: Food and Agricultural Organization of the United Nations.

Warren, D.M. 1991. 'Using Indigenous Knowledge in Agricultural Development'. *World Bank Discussion Paper No.127*. Washington, DC: The World Bank.

—— 1992. 'A Preliminary Analysis of Indigenous Soil Classification and Management Systems in Four Ecozones of Nigeria'. *Discussion Paper RCMD 92/1*, April 1992. Resource and Crop Management Division (RCMD), International Institute of Tropical Agriculture (IITA), Ibadan, Nigeria, and African Resource Centre for Indigenous Knowledge (ARCIK), Nigerian Institute of Social and Economic Research (NISER). Ibadan: IITA/ARCIK/NISER.

Warren, D.M., L. Jan Slikkerveen and S.O. Titilola (eds). 1989. 'Indigenous Knowledge Systems: Implications for agriculture and international development'. Center for Indigenous Knowledge for Agriculture and Rural Development. *Studies in Technology and Social Change No.11*. Technology and Social Change Program. Ames: Iowa State University.

Warren, D.M., L.J. Slikkerveer and D. Brokensha (eds). 1995. *The Cultural Dimension of Development: Indigenous Knowledge Systems*. Studies in Indigenous Knowledge and Development. London: Intermediate Technology Publications.

Warshall, P. 1989. 'Mali: Biological diversity assessment'. USAID, Bureau of Africa, Office of Technical Resources, Natural Resources Branch. Natural Resources Management Support Project, *Project No. 698–0467*. Tuscan: Office of Arid Lands Studies.

Waters-Bayer, A. and J. Farrington. 1993. 'Supporting Farmers' Research and Communication: The role of grassroots agricultural advisors'. *Quarterly of International Agriculture* Vol.32(2):170–87.

Watts, M. 1983. *Silent Violence: Food, famine & peasantry in Northern Nigeria*. Berkeley: University of California Press.

Weller-Molongua, C. and J. Knapp. 1995. 'Social Network Mapping'. *In* Slocum, Wichhart, Rocheleau and Thomas-Slayter (eds) *Power, Process and Participation: Tools for change*. London: Intermediate Technology Publications.

Wentling, M.G. 1983. '*Acacia albida*: Arboreal keystone of successful agro-pastoral systems in Soudano-Sahelian Africa'. Ithaca: Cornell University. *In* Seyler, J.R. 'A Systems Analysis of the Status and Potential of *Acacia albida* in the Peanut Basin of Senegal,'

Senegal Agricultural Research II Project. *AID Contract No. 685–0957–C-8004–00* (1993). Dakar: USAID.

Western, D. 1982. 'The Environment and Ecology of Pastoral People in Arid Savannas'. *Development and Change* Vol.13:183–211.

Wheeler, D. 1984. 'Sources of Stagnation in Sub-Saharan Africa'. *World Development* Vol.12(1):1–23.

Whyte, W.F. (ed). 1991. *Participatory Action Research*. Newbury Park: Sage Publications.

Wiggins, S. 1995. 'Changes in African Farming Systems Between the mid-1970s and the mid-1980s'. *World Development* Vol.7(6):807–48.

Wilding, L.P. and L.R. Drees. 1978. 'Spatial Variability: A pedologist's viewpoint'. *In* Stelly, M. (ed.) *Diversity of Soils in the Tropics*. American Society of Agronomy, *Special Publication Number 34*. Madison: ASA.

Willis, R., M. Suso, E. Kunjo and N.Y. Baldeh. 1995. 'Evaluation of the Community-Based Experimentation and Extension (CBEE) Project Implemented by the Association of Farmers Educators and Traders (AFET)'. Mimeo. Yundum: National Agricultural Research Institute.

Wilson, H.T. 1984. *Tradition and Innovation: The idea of civilization as culture and its significance*. London, Boston: Routledge & Kegan Paul.

Wilson, P. 1983. *Second-Hand Knowledge: An inquiry into cognitive authority*. Westport, Conn.: Greenwood Press.

World Bank. 1974. 'Appraisal of Integrated Rural Development Project, Mali'. Agriculture Projects Department. West Africa Regional Office. *Report No. 340a-MLI*. Washington, DC: World Bank.

—— 1979. 'Agricultural Sector Assessment for Nigeria'. Washington, DC: The World Bank. *In* Watts, M. *The Silent Violence: Food, Famine & Peasantry in Northern Nigeria*. Berkeley: University of California Press.

—— 1989. *Sub-Saharan Africa: From crisis to sustainable growth*. Washington, DC: World Bank.

—— 1990. *Vetiver Grass: The hedge against erosion*. 3rd edn. The Asia Technical Department, Agriculture Division. Washington, DC: World Bank.

—— 1995. *World Development Report 1995*. Washington, DC: World Bank.

—— 1996. *The World Bank Participation Source Book*. Washington, DC: World Bank.

—— 1997. *African Development Indicators*. Washington, DC: World Bank.

WRI (World Resources Institute). 1994. 'World Resources 1994–95: A guide to the global environment'. New York and Oxford: Oxford University Press.

Yeboah, A.K., J.S. Cadwell, M. Fofana and T.B. Maiga. 1991. *Utilisation de Critere de Classification Multiple pour l'Identification des domaines de Recommendation et de Recherche*. Bamako: DRSPR.

Yeboah, A.K. 1993. *Farming Systems Research and Extension Project, Mali: Intermediate Impact Indicators*. Bamako: DRSRP/USAID/SECID.

Zahan, D. 1960. *Sociétés d'Initiation Bambara: Le n'Domo, Le Korè*. Paris: Mouton & Co.

—— 1974. *The Bambara*. Leiden: E.J. Brill.

—— 1979. 'Principes de Médecine Bambara'. In Ademuwagun, Z.A., J.A.A. Ayoade, I.E. Harrison and D.M. Warren (eds) *African Therapeutic Systems*. Waltham: Crossroads Press.

Zaretzke, K. 1982. 'The Idea of Tradition'. *The Intercollegiate Review* Spring/Summer: 85–96.

Zolad. 1985. 'Projet de Recensement des Technologies Traditionnelles au Mali'. *Dossiers Techniques. Projet de Recensement des Technologies Traditionnelles*. Bamako: Division du Machinisme Agricole.

Government documents

Anon./OHVN 1992. *Etude du Système de Vulgarisation de l'OHVN*. DRAFT.

CNPER (Commission Nationale de Planification de l'Economie Rurale). 1972. *Rapport Final de la Commission Nationale de Planification de l'Economie Rurale. Pour l'elaboration de Plan Quinquennal* 1974–78. *Premiére Partie: Synthése des travaux des groupes d'études des zones rurales*. Bamako: CNPER.

Direction Générale du Plan et de la Statistique. 1972. *Evolution des Activites du Secteur Economique Organisé d'Aprés les resultats des Enquétes Auprés des Enterprises* 1960–1970. Bamako: Service de la Statistique Generale de la Comptabilite Nationale et de la Mechnograpie.

DNM (Direction Nationale de la Métérologie). 1992. *Rapport Technique de la Campagne Agricole* 1991–1992. Division de l'Agrométéorologie, Section Etudes et Développement. Bamako: DNM.

DRSPR (Département de Recherche sur les Systèmes de Production Rurale). 1987. *Commission Technique sur les Systèmes de Production. Document* No.5 *Volet OHV: Choix des Villages de Recherche; Enquête Typologie*. Bamako: DRSPR (Volet OHV).

—— 1988a. *Comité Technique Régional: sur les Systèmes de Production Rurale Centre de Sotuba, Résultats Campagne* 1987–88. Bamako: DRSPR (Volet OHV).

—— 1988b. *Commissions Techniques Spécialisées sur les Systèmes de Production Rurale: Résultats Campagne* 1987–88. Bamako: DRSPR (Volet OHV).

—— 1989. *Commissions Techniques Spécialisées sur les Systèmes de Production Rurale: Résultats Campagne* 1988–89. Bamako: DRSPR (Volet OHV).

—— 1990a. *Commissions Techniques Spécialisées sur les Systèmes de Production Rurale: Résultats Campagne* 1989–90. Bamako: DRSPR (Volet OHV).

—— 1990b. *Plan a Long Terme du Volet OHV du DRSPR: Programme de recherche* (1990–1994). Bamako: DRSPR (Volet OHV).

—— 1990c. *Cadre de Collaboration DRSPR-OHV*. Bamako: DRSPR (Volet OHV).

—— 1991a. *Comité Technique Régional: sur les Systèmes de Production Rurale Centre de Sotuba, Résultats Campagne* 1990–91. Bamako: DRSPR (Volet OHV).

—— 1991b. *Commission Technique Spécialisée sur les Systèmes de Production Rurale: Résultats Campagne* 1990–1991. Bamako: DRSPR (Volet OHV).

—— 1991c. *Approche Recherche Système: Note méthodologique* No.1. Sikasso: DRSPR (Sikasso).

—— 1992a. *Comité Technique Régional sur les Programme Systèmes de Production Rurale, Centre de Sotuba, Résultats Campagne* 1991–92. Bamako: DRSPR (Volet OHV).

—— 1992b. *Synthèse Evaluation Paysanne*. Bamako: DRSPR (Volet OHV).

—— 1992c. *Caracterisation Socio-Economique des Unites de Production en Zone de l'Office de la Haute Vallée du Niger*. Bamako: DRSPR (Volet OHV).

—— 1992d. *Information Générales sur les Villages de Recherche du DRPSR Sotuba et Taux d'Adoption des Technologies par les Paysan pour L'Ensemble de la Zone OHVN*. Bamako: DRSPR (Volet OHV).

—— 1992e. *Prévulgarisation Concept et Méthodologie: Note méthodologique* No.5. Sikasso: DRSPR (Sikasso).

—— 1992f. *Prise en Compte des Préoccupations des Femmes dans les Activités de Recherche du Volet OHVN: Resultats, Difficultes et Perspectives*. Bamako: DRPSR (Volet OHV).

—— 1992g. *Commission Technique sur les Systèmes de Production. Document* No.5, *Volet OHV, A. Choix des Villages de Recherche B. Enquete Typologie*. Bamako: DRSPR (Volet OHV).

—— 1993a. *Centre Regional de Recherche Agronomique de Sotuba*. 1993a. *Comité Technique Regional: Sous Programme Systemes de Production Rurale, Resultats Campagne* 1992–93. Bamako: DRPSR (Volet OHV)

—— 1993b. *Fiches Techniques*. Bamako: DRPSR (Volet OHV).

—— 1993c. *Synthese des Resultats* 1986–1992. Bamako: DRPSR (Volet OHV).

—— 1993d. *Synthèse Evaluation Paysanne*. Bamako: DRSPR (Volet OHV).

IER (Institut d'Economie Rurale). 1977. *Evaluation de 'Opération Arachide et Cultures Vivières: Analyse de exploitations agricules de l'OACV effectuée à partire de l'enquête descriptive de* 1976. Bamako: IER.

——1978. *Evaluation de l'Opération Arachide et Cultures Vivrières: Note de Synthèse Principaux résultats des Enquêtes Menées en* 1976 *et* 1977 *sur l'OACV*. Bamako: IER.

——1990. *Memento Techniques Culturales*. Bamako and Amsterdam: IER/DRSPR/CMDT/Institut Royal des Tropiques (KIT).

——1992. *Plan Strategique de la Recherche Agronomiue du Mali: Options de développement pour l'IER*. Bamako: IER.

MDRE Ministère du Développement Rural et de l'Environnement. 1997. Rapport d'Evaluation du Projet Développement de la Haute Vallée (DHV)(Projet No. 688–0233). Bamako: MDRE(CPS)/OHVN/USAID.

Ministère du Plan. 1987. *Esquisse du cadre d'un plan d'action de restructuration cas par cas des Opérations de Développement rural. Document provisoire*. Bamako: Min. du Plan.

Ministère du Plan et de l'Economomie Rural. (no date). *Organization de Monde Rural en République du Mali*. Action Rurale Édition Spéciale. Paris: Imprimerie Maubert et Cie.

Ministère du Plan et de la Statistique. (no date). *Plan Quinquennal de Développement Economique et Social*. Bamako: Min. du Plan et de la Statistique.

Ministère de la Production, Service de l'Agriculture. (no date). *L'Agriculture au Mali: Situation actuelle. Perspectives*. Bamako: Min. de la Production.

Ministère de la Production. 1972. *Etude Preparatoire a l'Execution de la Campagne de Vulgarisation de Masse pour la Protection des Semences et des Recoltes au Mali. Tome II: Etude des attitudes et motivations des élites paysannes. Contract d'Etude* No.861. Paris and Rome: MARCOMER and So.TE.S.A.

OHVN (Office de la Haute Vallée du Niger). (no date). *Termes de Reference des Spécialistes de la Division Vulgarisation*. Bamako: OHVN/Division Vulgarisation.

——1986. *Categorisation des Exploitations en Zone OHV: Séminaire de Oueléssébougou*. Bamako: OHVN/Division Stat. Plan. Evaluation.

——1988. *Projet de Restructuration de l'OHV*. Bamako: OHVN.

——1989. *Structure Organisationnelle et Attributions des Agents de la Division Vulgarisation*. Bamako: OHVN/Division Vulgarisation.

——1990. *Evaluation Activitiés des Chefs Sous Secteur*. Bamako: OHVN/DCDR.

——1991a. *Etat d'Avancement de l'Execution des Objectifs Assignés á la Vulgarisation dans le Cadre du Projet DHV*. Onzième Seminaire Annuel de l'OHV Tenu les 23–24–25 Avril, 1991, á Oueléssébougou. Bamako: OHVN/Division Vulgarisation.

——1991b. *Buts et Objectives du Projet*. Bamako: OHVN/Cellule de Suivi.

——1991c. *Structure Organisationnelle de l'Opération Haute Vallée*. Bamako: OHVN/Cellule de Suivi.

——1992a. *Rapport Annuel d'Activités: Campagne Agricole 1991–1992*. Bamako: OHVN.

——1992b. *Plan de Campagne Agricole: 1992–93*. Bamako: OHVN.

——1992c. *Communication OHVN au Seminaire sur la Recherche Systeme a Sikasso du 17–21 Novembre, 1992*. Bamako: OHVN.

——1992d. *Evaluation Paysans Pilotes*. Bamako: OHVN/DCDR.

——1993a. *Rapport Annuel d'Activités: Campagne Agricole 1992–1993*. Bamako: OHVN.

——1993b. *Plan de Campagne Agricole: 1993–94*. Bamako: OHVN.

——1993c. *Programme d'Activites P.N.V.A.: Campagne 1993/94*. Bamako: OHVN.

République du Mali. 1994. *Comptes Economiques du Mali: Series revisées 1980–1991. Resultats Provisoires* 1992–93–94; *Resultats Previsionnels* 1995. Direction Nationale de la Statistique et de l'Informatique: Bamako. *In* Chenevix-Trench *et al.* (1997), Land, Water and Local Governance in Mali: Rice production and resource use in the Sourou Valley, Bankass Cercle. Rural Resources Rural Livelihoods *Working Paper No. 6*. Institute for Development Policy and Management: University of Manchester.

SRCSS (Section de Reglementation et de Controle des Semences Selectionnées). 1987. *Catalogue Officiel des Espèces et Variétés Tome 1, Espèces Vivriéres*. Sotuba-Bamako: IER/DRA.

SRCVO (Section de Recherche sur les Cultures Vivrière et Oléagineuses). 1979. *Projet SAFGRAD au Mali: Rapport de la Campagne 1979*. Bamako: SRCVO.

——1989. *Projet Sol-Eau-Plante. Agroclimatologie Operationnelle. Resultats de la Campagne* 1988–89 *et propositions de programme* 1989. Commission Technique Spécialisée des Productions Vivrières et Oléagineuses. Bamako du 11 au 15 Avril 1989. Bamako: SRCVO.

——1992. *Projet Sol Eau Plante: Agroclimatologie Opérationnelle. Résultats de la Campagne* 1991–1992, *Propositions de Programmes* 1992. *Comité du Programme Ressources Naturelles*. Bamako: SRCVO.

Union Soudanaise RDA (no date). *La Politique Economique de l'Union Soudanaise RDA*.